The Practical Craft

Readings for Business and Technical Writers

F14.20

The Practical Craft

Readings for
Business and Technical Writers

W. Keats Sparrow / *East Carolina University*

Donald H. Cunningham / *Morehead State University*

Houghton Mifflin Company / *Boston*

Dallas   Geneva, Illinois   Hopewell, New Jersey   Palo Alto   London

"Analyze the 'What' Before You Write the 'How to'" by Forrest G. Allen reprinted from the January 1974 issue of *Industrial Education* magazine with permission of the publisher. This article is copyrighted. © 1974 by Macmillan Professional Magazines, Inc. All rights reserved.

Printed in the U.S.A.

Library of Congress Catalog Card Number: 77-93967

ISBN: 0-395-25590-2

To Nicole and Mark

Contents

PART III

What are some important writing strategies?

PART IV

What are some important types of letters
and reports?

PART V

What are the important formal elements
of reports?

Preface

If your work requires you to write, this book is for you. If you are a two-year or four-year college student in business or economics, in engineering or technology, or in physical, biological, or social science, this book is for you. If you are already a business person, administrator, lawyer, doctor, scientist, engineer, technician, or social worker, this book is for you, too.

Of course, of the many books on business and technical writing available to you, if the one you choose as a guide is adequate it at least touches on the topics covered by this book. But because of limitations of space and author perspective, rarely does a book say all that needs to be said about such important topics for business and technical writers as style, strategy, or technique. And therein lies the difference between this book and others on the same subject: this book brings together for you in a convenient volume, from a variety of sources and perspectives, some of the most serviceable information you can find about writing for business and industry. It thus affords you a wide range of views about several of the most important concerns of business and technical writers.

Arranged in five parts, the book is designed specifically to help you answer the what, why, and how of business and technical writing. Part I considers the questions of what is business and technical writing and why study such writing. The other four parts consider some major problems of how to write effectively for business and industry. Part II helps you develop an appropriate writing style; Part III offers you several writing strategies; Part IV presents you with solutions for several common but difficult writing assignments; and Part V explains for you the formalities of report writing. The introduction to each part provides you with a brief perspective on the topic covered in that part and directs your attention to the important concepts in the selections that follow. And the annotated bibliography at the end of each part will lead you to further readings on the topic discussed in that part.

Yet, in spite of its inclusiveness, this book is not a panacea for business and technical writers. It makes no claim, for example, at including all of the many excellent articles that exist on business and technical writing. And it makes no attempt to provide models or to answer questions about usage— for those matters you will still need your writer's handbook and other reference books. Moreover, you will find that the selections range from the general to the specific and that some selections bear more directly than others on the writing assignments for which you are responsible.

But in each part of the book you will find a number of selections that broaden your perspective about the kind of writing you encounter as a part of your day-to-day work. And by helping you understand what kinds of writing are required of you, why a mastery of writing is important to you, and how to improve the writing done by you, this book should ease the misgivings you may have over the writing responsibilities you face.

If so, it will have accomplished its purpose.

W. K. S.
D. H. C.

The Practical Craft

Readings for Business and Technical Writers

Part I

What Is Business and Technical Writing and Why Study Such Writing?

"A technical writer might write instructions on how to program a computer, or how to maintain it, or how to repair it. The business writer would be more likely to write instructions on how to schedule work for the computer, how to get in line for computer time, how to budget the costs of computer time."

Francis W. Weeks*

"As an employee you work with and through other people. This means that your success as an employee—and I am talking of much more here than getting promoted—will depend on your ability to communicate with people and to present your own thoughts and ideas to them so they will both understand what you are driving at and be persuaded. The letter, the report or memorandum, the ten-minute spoken 'presentation' to a committee are basic tools of the employee."

Peter F. Drucker†

The writing for which you are responsible in business and industry is similar in most respects to the writing of newspaper reporters, of speech writers, of historians, of novelists, and of anyone else whose success depends largely on ability with language. Whether writing a report or a novel, for example, you must have a command of punctuation, mechanics, grammar, and spelling—or else you might confuse your reader with inaccuracies and project an untutored image of yourself. From a flexible vo-

---

* Executive Director, American Business Communication Association. (By permission)
† From "How to Be an Employee," *Fortune*, May 1952, p. 126. (Reprinted from *Fortune Magazine* by permission)

cabulary, you must be able to choose words whose denotations and connotations communicate your meaning with precision, and you must be able to fit those words together into clear sentences. You must also be able to arrange your sentences into coherent paragraphs and then be able to arrange your paragraphs into sensible sequences.

Yet in order to be a successful writer in business and industry, you must have a command of many special writing skills that are not ordinarily taught in conventional writing courses or explained in traditional handbooks. For instance, you must have a thorough familiarity with the formats of letters, memorandums, reports, and other organizational patterns that are standard in business and industry. You must know how to use such visual aids as itemizations, captions, and graphs in order for you to communicate technical information with the utmost clarity. You must become unusually shrewd at anticipating your readers' reading abilities and special needs so that you can accommodate those factors to advantage in both how you write and what you write. And you must learn to exercise careful judgment when such conventional writing "rules" as those about variation in sentence structure, lengthy paragraph development, or inductive ordering of material might prove counterproductive to your purpose in writing.

Common names for college or adult education courses that teach the special skills you need to write for business and industry are, to name a few, Business Writing, Technical Writing, Report Writing, Scientific Writing, and Writing for Business and Industry. Under whatever names, such courses usually overlap in content in that what you learn from them is essentially the same—how to write (and speak) about your professional concerns in clear, concise, purposeful English. But the courses and the writing done in them differ in emphasis. And, as suggested by the various course names mentioned here, the differences result largely from the needs of the students enrolled in them.

For example, a course in business writing would stress reader motivation in sales letters more than would courses in technical or scientific writing. A course in technical writing would place more emphasis on how to write detailed descriptions or instructions than would courses in business or scientific writing. And a course in scientific writing would give closer attention to drawing inferences from data and to writing summaries of laboratory experiments than would courses in business and technical writing. A course with a general name like Writing for Business and Industry might be designed to accommodate a diverse group of students from business, technology, and science.

Whether you are in business, industry, or science, your work will indeed require you to write and speak. And the more successful you are in your field of specialization—that is, the more you advance in your profession—the more you will be required to write and speak. You will have to write letters, memorandums, proposals, reports; and, if you are a specialist, you may have to write descriptions, manuals, laboratory analyses, or a number of other types of technical communications. You will also have to talk—to speak on the telephone or face-to-face with clients and coworkers, to give directions or orders, to explain technical matters, and perhaps to give an occasional oral presentation.

And for everything you write and say, you will be judged. If you use standard English (the English taught in classrooms and found in handbooks), and if you communicate your ideas with clarity, conciseness, and grace, you and what you say will be judged in a favorable light—by your superiors, by your peers, by your subordinates, and by the public. But if you are unable to communicate clearly in standard English, you and what you say will be judged in an unfavorable light, thus diminishing your chances for success.

You may think you can resolve your writing chores quite simply by turning them over to an editor or "ghost writer"—your secretary, perhaps, or the professional writers in the publications and presentations department. A few people in the workaday world are fortunate enough to have recourse to a competent writing assistant; but more often than not such is not the case.

In the first place, as a professional in your area of specialization, you probably have a better preparation in writing and speaking than your secretary (whose training, after all, is usually in typing rather than in writing). In the second place, only a handful of companies can afford to maintain a staff of professional writers to whom other employees can turn for assistance; and the few staffs of professional writers to be found probably have assigned tasks—editing a company newsletter, preparing advertisements, assisting chief executives—that prevent them from being of much help to you. Moreover, even if you should have the services of a professional writer, from time to time you will find that early deadlines or the absence of a readily available writer will still require you to do your own writing.

The only workable solution, then, is to dispel your fantasies about ghost writers and become what a successful professional in business, industry, government, law, or science is required to become—a competent amateur writer.

Although none of them attempts an inclusive definition, the first three essays in Part I help you see more clearly what is meant by the terms "business writing" and "technical writing." The second three essays explain why and in what circumstances your work demands that you have a mastery of communication skills.

You will learn from reading the essays that your ability to write and speak well is an ability that far more people than English teachers consider important for success in your career.

Selection One

Some Principles of Business Communication

ROBERT L. SHURTER AND J. PETER WILLIAMSON

> *When they wrote the following article, Robert Shurter was an English professor at Case Institute of Technology and Peter Williamson was a business administration professor at Dartmouth College. Their explanation of business communication should help you to understand the writing problems you face in business and industry.*

Much of the world's business would end tomorrow if we paid more than lip service to the maxims we have inherited on the subject of communication. At one time or another, all of us have glibly repeated such old saws as "Silence is golden," "No news is good news," "What you don't know won't hurt you," and have expressed admiration for small boys who are "seen and not heard," perhaps because they will grow up to be "the strong, silent men" considered heroic in certain types of fiction.

These half-truths of our folklore are nonsense in today's complex, specialized business community, where a continuous exchange of ideas and informaton is an absolute necessity. "The first executive function," says Chester I. Barnard in his definitive book *The Functions of the Executive*, "is to develop and maintain a system of communication."* In the same vein, the editors of *Fortune* recently commented, "If business has a new motto, 'Communicate or Founder' would seem to be it."†

From *Written Communication Is Business* by Shurter & Williamson, 3rd ed. Copyright © 1971 by McGraw-Hill, Inc. Used with permission of McGraw-Hill Book Company.
* Reprinted by permission of the Harvard University Press.
† Reprinted by special permission from *Fortune's* Communication Series, *Is Anybody Listening?* copyright 1950, by Time, Inc.

These comments should suffice to show the importance of communication skills in business. What do we mean by "communication"? We can define it quite simply as *imparting or exchanging thoughts or information*—and since we are dealing only with one form of communication, we must add *in writing*. But such a definition really doesn't help very much. We can get closer to the fundamentals of written communication by thinking of it as a process which always includes:

1. A writer.

2. The material—facts, ideas, information, recommendations, conclusions—which he wants to communicate.

3. A reader. For certain forms of business communication we would be more realistic to say "a group of readers."

This analysis may be oversimplified, but it is better than the abstractions of our first definition because it puts human beings—a writer and reader(s)—into the act of communication. "It takes two to speak the truth," said Thoreau, "—one to speak and another to hear."

But in this two-person situation let there be no doubt about who bears the responsibility for effective communication. The responsibility rests on you, as the writer. You might as well accept this responsibility right now. It will be forced on you in business. You will have to abandon certain alibis you may have used in the past. You won't be able to blame misunderstandings on "a stupid reader"; you'll have to make every effort to write with such clarity and simplicity that he can understand. You can't say that he is stubborn or pigheaded or narrow-minded because he doesn't agree with you; you'll have to use tact and persuasion and evidence to make him see your point of view. Of course, you may still fail to get your ideas across; it would be unrealistic to think that you can always succeed with your reader. But if you fail, you have at least done so with the knowledge that you did your best—and that is the essence of responsibility in writing, as in anything else.

Fulfilling your responsibility requires that you *think*. Think *before* you write, *when* you write—and then think about how you can improve or revise *after* you have written. Writing which serves your particular purpose requires that you think the purpose through. Clear writing stems from thoughtful planning. Concise writing results from thinking your way through to essentials, eliminating the extraneous and irrelevant. And writing which is correct and appropriate in style reveals that you have thought of how the reader will react and have designed your communication to produce the reactions you want.

We can sum up these observations on thinking . . . by saying that you must have your reader always in mind. You must always be thinking of what *he* wants to learn, what you want *him* to learn, what reactions you want to produce in *him* and that you want to avoid, and how your writing can accomplish this.

The more you can learn about orderly habits of thought and the logical sequence of events and ideas, the better you can organize your material. The more you can learn about psychology and human relations and people, the more you can know about your readers. These habits of thought and a broad knowledge of people are the most useful background you can have for writing in business. They will enable you to avoid the pitfalls of "thought-less" writing and accomplish the particular purposes you have in mind for a communication. You will, of course, have to supplement this background with a knowledge of the techniques of writing . . . as they apply to the specific problems of writing letters, reports, memorandums, and other forms of business communication. But whatever the form, the fundamentals remain the same.

When business writers forget the fundamentals of the communication process and ignore the thinking behind it, a phenomenon known to electrical engineers as *noise* or *interference* occurs. In their diagrams of mechanical or electronic communication, the engineers show messages from source to destination like this:

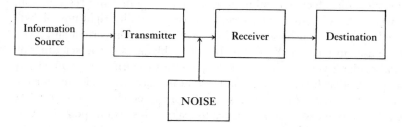

The noise here is static on your radio, "snow" or fading on your television screen—in short, anything which blocks or impedes the transmission of sound or sight. Unfortunately, human beings are probably even more ingenious when it comes to producing noise or interference in their written communications. The noise caused by thoughtlessness—whether in the form of ambiguity, tedious wordiness, inappropriate style, distracting irrelevancies, or any of a number of writing faults—damages the impression you are trying to plant in your reader's mind.

For the most part, good business writing is just good writing. But if your education until now has involved only nonbusiness writing—essays and short stories, for example—you may be disappointed by your first attempts at business writing. We . . . stress the importance in business writing of having your purpose clearly in mind, and this, of course, applies to all writing. If you really understand the elements of good writing and are aware of the particular purposes they serve, then you will be successful whether you are writing an essay, short story, magazine article, scholarly article, or a business letter or report. But if what you have learned is how to write a good essay or short story without understanding the techniques involved and knowing *why* they are suitable, then you will have trouble with business writing, just as a person trained to write business communications without understanding the *why* of what he is doing will have trouble with essays and short stories.

Some distinctions between business and literary writing are quite apparent. The reader of a business communication is generally in much more of a hurry than is the reader of a literary work. Conciseness, brevity, and devices such as headings and summary sentences or paragraphs are more important to business than to literary writing. At the same time, the reader of a business communication is rarely looking for entertainment. Witticisms, clever allusions, and paradoxes may add much to the enjoyment of literary writing, but they are more apt to annoy the business reader. . . .

One aspect of business writing that you can learn only on the job concerns the particular forms, procedures, and even writing style that have been established by the business organization you work for. Many companies have a single acceptable form for letters, for example. Some companies insist that letters and reports conform to a particular style manual, such as the *New York Times Style Book*. These details should not concern you now. It will be easy to adapt your writing to them if you understand the fundamentals of good business writing and the purposes they serve.

DISCUSSION AND ACTIVITY

1. Shurter and Williamson mention several writing faults that lead to "noise" or "interference" in written communications. Name those writing faults and explain how each could interfere with the effective communication of ideas.

2. Explain why the reader of a business communication might be in a greater hurry than the reader of a literary work. Why is the reader of a business communication not looking for entertainment?

Selection Two

What Is Technical Writing?

W. EARL BRITTON

W. Earl Britton is a professor emeritus of the department of humanities, College of Engineering, the University of Michigan. His definition of technical writing given here extends beyond the older definitions in the first part of his article.

Although the dean of an engineering college once denied the very existence of technical writing, many of us are confident of its reality. But we are not sure that we can convince others of its uniqueness. This uncertainty deepens when we observe the variety of activities incorporated under this label, as well as those that barely elude its scope. Our schools do little to clarify the situation with such course titles as technical writing, engineering writing, engineering English, scientific English, scientific communications, and report writing. Nor do the national societies help with their emphasis upon medical writing, biological writing, science writing, and business English. Applying the general term *technical writing* to a field of such diversified activities is convenient but misleading, yet this is the current practice. In view of the confusion, there is little wonder that a teacher in this field should often be asked, even by colleagues, "What is this technical writing you teach, and how does it differ from any other?"

In addition to satisfying this query, a truly helpful definition should go much further and illuminate the tasks of both the teachers and authors of technical writing. This requirement has been fulfilled in varying degrees by a number of definitions already advanced, the most significant of which form four categories.

Technical writing is most commonly defined by its subject matter. Blickle and Passe say:

Any attempt . . . to define technical writing is complicated by the recognition that exposition is often creative. Because technical writing often employs some of the devices of imaginative writing, a broad definiton is necessary. Defined broadly, technical writing is that writing which deals with subject matter in science, engineering, and business.[1]

Mills and Walter likewise note that technical writing is "concerned with technical subject matter," but admit the difficulty of saying precisely what a technical subject is. For their own purpose in writing a textbook, they call a technical subject one that falls within science and engineering. They elaborate their view by adding four large characteristics of the form: namely, its concern with scientific and technical matters, its use of a scientific vocabulary and conventional report forms, its commitment to objectivity and accuracy, and the complexity of its task, involving descriptions, classifications, and even more intricate problems.[2]

The second approach is linguistic, as illustrated in an article by Robert Hays, who, admitting the existence of technical writing without actually defining it, remarks upon its conservativeness, its Teutonic subject-verb-object word order, and the fact that it shares with other forms of writing the "common" English vocabulary. However, he cites two fundamental differences between technical and other prose. The psychological difference is the writer's "attitude of utter seriousness" toward his subject, and his dedication to facts and strict objectivity. But the greater difference, at least for "teachers and students of technical writing, is linguistic," in that technical style demands a "specialized vocabulary, especially in its adjectives and nouns."[3]

The third definition concentrates on the type of thought process involved. This approach underlies some of the research being directed by A. J. Kirkman, of the Welsh College of Advanced Technology in Cardiff, who has been investigating the causes of unsatisfactory scientific and technical writing. His group is examining in particular the suggestion that a distinction exists between ways of thinking and writing about literary and scientific subjects. The theory postulates two types of thinking, each with its own mode of expression. Associative thought belongs to history, literature, and the arts. Statements are linked together by connectives like *then* and *rather*, indicating chronological, spatial, or emotional relationships. Sequential thought belongs to mathematics and science. Statements are connected by words like *because* and *therefore*, revealing a tightly logical sequence. Professor Kirkman suggests that the weakness of much scientific writing results from forcing upon scientific material the mode of expression appropriate to the arts. He adds:

The important distinction is that sequential contexts call for comparatively inflexible lines of thought and rigid, impersonal forms of expression, whereas associative contexts permit random and diverse patterns of thought which can be variously expressed.[4]

Finally, technical writing is sometimes defined by its purpose. This approach rests upon the familiar differentiation between imaginative and expository prose, between DeQuincey's literature of power and literature of knowledge. Brooks and Warren find the primary advantage of the scientific statement to be that of "absolute precision." They contend that literature in general also represents a "specialization of language for the purpose of precision" but add that it "aims at treating kinds of material different from those of science," particularly attitudes, feelings, and interpretations.[5]

Reginald Kapp pursues a similar line by dividing writing into imaginative and functional literature. Imaginative literature involves personal response and is evocative; functional literature concerns the outer world that all can see. Functional English, he says, presents "all kinds of facts, of inferences, arguments, ideas, lines of reasoning. Their essential feature is that they are new to the person addressed." If imaginative literature attempts to control men's souls, functional English should control their minds. As he writes, the technical and scientific author "confers on the words the power to make those who read think as he wills it."[6]

All of these approaches are significant and useful. Kirkman's suggestions are certainly intriguing, but I find Kapp's classification particularly helpful and want to extend it slightly.

I should like to propose that the primary, though certainly not the sole, characteristic of technical and scientific writing lies in the effort of the author to convey one meaning and only one meaning in what he says. That one meaning must be sharp, clear, precise. And the reader must be given no choice of meanings; he must not be allowed to interpret a passage in any way but that intended by the writer. Insofar as the reader may derive more than one meaning from a passage, technical writing is bad; insofar as he can derive only one meaning from the writing, it is good.

Imaginative writing—and I choose it because it offers the sharpest contrast—can be just the opposite. There is no necessity that a poem or play convey identical meanings to all readers, although it may. Nor need a poem or play have multiple meanings. The fact remains, nevertheless, that a work of literature may mean different things to different readers, even at different times. Flaubert's *Madame Bovary* has been interpreted by some as an attack on romanticism, and in just the opposite way by others who read it as an attack on realism. Yet no one seems to think the less of the novel. Makers of the recent film of *Tom Jones* saw in the novel a bedroom farce, whereas serious students of Fielding have always regarded it as an effort to render goodness attractive. Varied interpretations of a work of literature may add to its universality, whereas more than one interpretation of a piece of scientific and technical writing would render it useless.

When we enter the world of pure symbol, the difference between the two kinds of communication—scientific and aesthetic—becomes more pronounced. Technical and scientific writing can be likened to a bugle call, imaginative literature to a symphony. The bugle call conveys a precise message: get up, come to mess, retire. And all for whom it is blown derive identical meanings. It can mean only what was intended. But a symphony, whatever the intention of the composer, will mean different things to different listeners, at different times, and especially as directed by different conductors. A precise meaning is essential and indispensable in a bugle call; it is not necessarily even desirable in a symphony.

The analogy can be extended. Even though the bugle call is a precise communication, it can be sounded in a variety of styles. An able musician can play taps with such feeling as to induce tears, and it is conceivable that a magician might blow reveille so as to awaken us in a spirit of gladness. The fact that scientific writing is designed to convey precisely and economically a single meaning does not require that its style be flat and drab. Even objectivity and detachment can be made attractive.

Because technical writing endeavors to convey just one meaning, its success, unlike that of imaginative literature, is measurable. As far as I am aware, there is no means of determining precisely the effects of a poem or a symphony; but scientific analyses and descriptions, instructions, and accounts of investigations quickly reveal any communication faults by the inability of the reader to comprehend and carry on.

Objection may be raised to this distinction between the two kinds of writing because it makes for such large and broad divisions. This I readily admit, at the same time that I hold this feature to be a decided advantage, in that it removes the difficulty that usually arises when technical writing is defined by its subject matter.

Emphasis upon engineering subject matter in technical writing, for example, has implied that engineering has a monopoly on the form, and that a Ph.D. dissertation in linguistics or even certain kinds of literary criticism and a study of federal economic policy are other kinds of writing. When all such endeavors that convey single meanings are grouped under the label technical and scientific writing, or some other term, for that matter, then division of these into subject areas, instead of creating confusion, becomes meaningful. Some subjects will be far more technological than others, ranging from dietetics to nuclear fission, and in some instances being related to science only by method of approach; some subjects will offer more linguisitic difficulty than others; some will require a tighter, more sequential mode of thinking; but all will have in common the essential effort to limit the reader to one interpretation.

It seems to me that this view not only illuminates the nature of technical writing but also emphasizes the kind of training required of our schools.

Unfortunately, few educational institutions are meeting the needs in this field. Professor Kirkman mentions the failure of the traditional teachers to provide enough practice in writing on practical subjects. Professor Kapp has insisted that the conventional instruction in formal English courses does not equip a man to teach or practice scientific and technical writing. The Shakespearian scholar, G. B. Harrison, commenting upon his formal writing courses in England, says:

The most effective elementary training I ever received was not from masters at school but in composing daily orders and instructions as staff captain in charge of the administration of seventy-two miscellaneous military units. It is far easier to discuss Hamlet's complexes than to write orders which ensure that five working parties from five different units arrive at the right place at the right time equipped with proper tools for the job. *One learns that the most seemingly simple statement can bear two meanings* and that when instructions are misunderstood the fault usually lies with the wording of the original order.[7] [My italics]

But our strictures should not be confined to the English teachers. All of us in education must share the responsibility for this condition. In fact, I believe that in all too many instances, at least in college, the student writes the wrong thing, for the wrong reason, to the wrong person, who evaluates it on the wrong basis. That is, he writes about a subject he is not thoroughly informed upon, in order to exhibit his knowledge rather than explain something the reader does not understand, and he writes to a professor who already knows more than he does about the matter and who evaluates the paper, not in terms of what he has derived, but in terms of what he thinks the writer knows. In every respect, this is the converse of what happens in professional life, where the writer is the authority; he writes to transmit new or unfamiliar information to someone who does not know but needs to, and who evaluates the paper in terms of what he derives and understands.

B. C. Brookes takes a similar position when he suggests that English teachers concerned with science students should ask them occasionally to explain aspects of their work which they know well so that the teacher who is acquainted with the material will understand it. Such an assignment not only is a real exercise in composition but also taxes the imagination of the student in devising illuminating analogies for effective communication. The teacher's theme should be: "If your paper is not plain and logical to me, then it is not good *science*."[8]

Both Harrison and Brookes recommend the kinds of exercises that are often viewed skeptically at the college level. This is a regrettable attitude, especially since it usually derives from unfamiliarity with the nature and need of such work and from unawareness of its difficulty and challenge. Teachers oriented primarily toward literature see little of interest in this field, but those who enjoy composition—especially its communicative aspect—can find considerable satisfaction here. Of one thing they can always be sure: deep gratitude from those they help.

NOTES

1. *Readings for Technical Writers*, ed. Margaret D. Blickle and Martha E. Passe (New York, 1963), p. 3.
2. Gordon H. Mills and John A. Walter, *Technical Writing* (New York, 1954), pp. 3-5.
3. Robert Hays, "What Is Technical Writing?" *Word Study*, April, 1961, p. 2.
4. A. J. Kirkman, "The Communication of Technical Thought," *The Chartered Mechanical Engineer*, December, 1963, p. [2].
5. Cleanth Brooks and Robert Penn Warren, *Understanding Poetry*, 3rd ed. (New York, 1960), pp. 4-5.
6. Reginald O. Kapp, *The Presentation of Technical Information* (New York, 1957), chaps. I-II.
7. G. B. Harrison, *Profession of English* (New York, 1962), p. 149.
8. B. C. Brookes, "The Teaching of English to Scientists and Engineers," *The Teaching of English*, Studies in Communication 3 (London, 1959), pp. 146-7.

DISCUSSION AND ACTIVITY

1. Britton mentions four approaches to defining technical writing. What are the four approaches? To phrase the question in a different way, in what ways is technical writing different from conventional writing?

2. In what respects is technical writing similar to business writing? In what respects do business writing and technical writing differ?

3. Read Robert Frost's poem "The Road Not Taken" and discuss its meaning with other members of your class. Select a passage that someone construes to have a meaning different from the one you see and discuss why the passage justifies both interpretations.

Selection Three

The Three-Horned Dilemma: Technical Writing, Business Writing, and Journalism

PAUL M. ZALL

Paul M. Zall is an English professor at California State University, Los Angeles. In the article that follows, he points up several distinctions and similarities among technical writing, business writing, and journalism.

Everyone knows a dilemma has only two horns. But since World War II, teachers of writing have been riding a chrome-plated, high-finned, three-horned dilemma, with no filter feedback. It used to be that they could get along just fine with one regular-sized course in report writing, or pre-professional writing, or just plain writing, but those days are gone forever. Nowadays, unless it's got a specialized tag to it—like business writing, technical writing, journalism—it just won't do. Is this progress?

This is to be expected, I suppose, in our crazy, mixed-up, transistorized, magnetohydrodynamic-powered world. Everybody wants to be a specialist. And as we open up new worlds we uncover new specialists. Since everybody wants to be a specialist, everybody wants to get an education—or at least a diploma.

DEMAND FOR SPECIALIZATION

And so we are stuck with our fractionated writing courses simply because we have this built-in demand for specialization. Tech writing, business writing,

From the *American Business Writing Association Bulletin*, March 1960. Reprinted with permission of the *Bulletin* of the American Business Communication Association.

journalism, only look like the real dilemma. The real dilemma is formed by pressure from the community to specialize, and by pressure from critics among ourselves urging us not to do what we have to do—namely, specialize.

The new standard procedure is to run a fund drive, build a new campus, hire Ph.D.'s with the ink still wet behind their ears, and send them out to survey business, industry, and government agencies to find out what they want us to teach. We then revise our ideas of curriculum to accommodate them. The result is: we have ultra-modern campuses, pin-striped professors, the finest in modern facilities and equipment—but neither time, energy, people, nor money to serve our proper customers, the students looking for an education.

We are bending so far over backwards to serve the needs of the community we are falling flat on our faculties. Naturally, if you ask a specialist what he wants in the way of education he will tell you, "More specialization." We are now running so many cafeteria courses in tech writing (elementary, advanced, king-size), business writing (ditto), and journalism (ditto), some professions are beginning to feel that if we do not give them equal time, we are insulting their professional status. Last fall we received a serious request to offer a course in writing for plumbers! We flushed it, thank heavens, when no one enrolled. The fact that no one enrolled shows there is still hope.

WHAT IS EXPECTED OF THE TECHNICAL WRITER

My own speciality is technical writing, though I prefer to call it industrial writing. It is changing so fast, we really do not know what to call it anymore. Not so long ago, it was addressed to a narrow, technical audience interested in technical subject matter. We used to get along with a few printed forms and simply filled in appropriate blanks to suit particular purposes.

Nowadays, however, a first-class industrial writer is expected to write articles, brochures, correspondence, film scripts, manuals, text books, talks, press releases, and almost anything else we can classify as writing. He is expected to be a jack-of-all-trades and a master of science to boot. He is expected to know something about astronomy, astrophysics, astrodynamics—everything from Aardvarks to Zymogeny.

And yet, what does the community want? I quote from a letter dated 15 October 1959 suggesting college courses in the area of industrial writing which would be taken *in addition to* the usual courses taken by students preparing for engineering degrees:

• *Freshman Year:* No special courses.

• *Sophomore Year:* Technical Writing as a Specialty (1 course), covering tech writing as compared to other types of writing. Second semester, another course by the same name, covering intensive study of the wide variety of documentation requirements for commercial or government customers.

• *Junior Year:* Preparation of Instruction Handbooks; second semester— Preparation of Instruction Manuals.

• *Senior Year:* Preparation of technical reports, specifications, proposals, bulletins, brochures, etc. Second semester—Preparation of a comprehensive technical report and preparation of a detailed technical proposal.

The letter closes with this comment, which will be allowed to speak for itself: "Courses in journalism and business administration would also be desirable to offer depth of experience."

This is not a particularly unusual request. What these writers do not realize is that we have had to fight for years to persuade the Engineering Department that *one* course in tech writing would not take too big a bite out of their technical empire. In fact, the breakthrough came because there is a conspicuous trend in our engineering schools away from such fractionated curricula as chemical engineering, electrical engineering, mechanical engineering, and so on, to a more comprehensive program they call Engineering Sciences, including at least one-fifth of the total program in Humanities and English.

Best of all, this trend in the Engineering schools is running counter to industry's demands for more specialized specialists. It is a case of *not* giving the Community what it wants but what's *good* for it. And yet, the squeeze on tech writing teachers increases. The technology in industry has advanced so far and so fast in the past few years that even engineers sitting side by side are actually beside themselves trying to understand what the other man is talking about. Take the new field called molecular electronics. Reading about it in the technical journals is an exercise in Scrabble: On Monday it's called "molectronics." On Tuesday, "semi-conductor solid-circuitry." On Wednesday, "integrated microcircuitry." On Thursday, "intrinsic microcircuitry." On Friday, "intrinsic solid-circuitry," and on Saturday, "electrinsic circuitry." Thank heavens for Sunday or they'd be back to "electricity."

It used to be that the only trouble was getting engineers to communicate with management, but now they have a problem communicating with themselves and with everyone else. Industrial activity is underwritten by tax dollars or consumer credit, so it has to be translated for the housewife in Peoria or the dowager in Duluth. The competition for public approval is outrageous—and gaining this approval involves writing, writing, and more writing, to all levels of American life. It used to be that industry could hire public relations people or advertising copy writers to do this job, but now the only people who can tell what is going on are the people who are making it go.

NO LONGER POSSIBLE TO TEACH FORMS OF WRITING

What that means is: we cannot be satisfied to teach the specialized forms of technical writing anymore. There can never again be a good course in what every tech writer should know. The field has become so big even in this specialty that we must leave it to each individual to pick up the specialized forms for himself. About all that we can hope to accomplish—and accomplish well—is to give our students something to say and show them *some* of the possible ways to say it.

For many reasons, our young people have nothing to say. So the first thing we do in our tech writing course is to make them find something worth saying. We set up a dummy company with a dummy President, Argle D. Bargle, and turn them loose on an actual project where we know little literature has been published and we know the only way they can find out anything is through actual correspondence and actual interview. We know they will run into contradictory evidence and run up against blank walls, and into dark alleys and security restrictions. We even change policy and procedures on them the day before an assignment is due—we do our utmost to create an actual industrial climate.

The idea of this play acting is to give them a broad range of experience in situations similar to the ones they are likely to encounter in real life. But the underlying intention is to make them so familiar with the climate of professional life that they will forget worrying about the forms of things in favor of acquiring the ability to organize order out of chaos, to see connections where no connections seem to exist, to slice through to the heart of a problem, to weigh alternative approaches and choose solutions with wisdom rather than out of expediency. In short we try to make them think. This wisdom can come only with broad experience, a feeling of familiarity, a sense of having been here before.

We do not have to be buffaloed by the pressure for specialized courses in the *forms* of things: proposals, analytical reports, memos, dunning letters, etc. These will be taken care of by company policy and procedure anyway, so we are not depriving our students of anything of value, if we should forget to include them in the course of a semester. Our concern ought to be the writing that will be poured into the forms.

WRITING IS WRITING?

And here is where we face the real three-horned dilemma. What it amounts to is this: There are those who insist that there is no real difference between tech writing, business writing, journalism, and so we ought not to be specializing at all. In fact, there are those who go even further and say all writing is alike: Writing is Writing. And they include literature, too. O.K., here's a piece of literature, familiar to one and all—Humpty Dumpty. Now here it is as it must be written by a journalist anxious to hold his job:

Humpty Dumpty sat on a wall,
Humpty Dumpty had a great fall;
All the King's horses and all the King's men
Couldn't put Humpty together again
Because he was an egg.

And now here it is as by a business writer:

Humpty Dumpty sat on a wall,
Humpty Dumpty had a great fall;
All the King's horses and all the King's men
Couldn't put Humpty together again
Until they got Gripsit Household Cement. . . . Got yours?

And finally, the same theme by a tech writer:

A 72-gram brown Rhode Island Red country-fresh candled egg was secured and washed free of feathers, blood, dirt, and grit. Held between thumb and index finger, about 3 ft. or more from an electric fan (GE Model No. MC-2404 Serial number JC23023, non-oscillating, rotating on "Hi" speed at approximately 105.23 plus or minus 0.02 rpm), the egg was suspended on a pendulum (string) so that it arrived at the fan with essentially zero velocity normal to the fan rotation plane. The product adhered strongly to the walls and ceiling and was difficult to recover. However, using putty knives a total of 13 grams was obtained and put in a skillet with 11.2 grams of hickory smoked Armours old style bacon and heated over a low Bunsen flame for 7 min. 32 sec. What there was of it was of excellent quality.
("The DP Report," Dupont Explosive Dept., Atomic Energy Div., Savannah River Laboratories, July 12, 1954.)

Writing is Writing? Anyone who knows anything at all about writing knows that imaginative literature is intentionally ambiguous—the more it leaves to the reader's imagination the better it is. It works by metaphor, and exploits the many meanings a word may have. Business writing may work this way, too, sometimes. It conscientiously avoids words like "flabby" and "sweat" because of their bad connotation—and uses words like "fully developed" and "perspiration." But business writing, like tech writing and journalism, cannot afford to leave itself open to the vagaries of the reader's imagination. The point is: business writing, journalism, technical writing are not literature. They are written to tell and to sell—seldom to entertain.

DIFFERENCES IN THESE FIELDS OF WRITING

Furthermore, each of the three fields requires its own language. Each requires its own words and its own music, too. We do not expect the businessman to talk like an engineer, or the engineer to talk like a journalist. Their languages are different. Their thinking is different. Each requires a different writing sense. By writing sense, I simply mean a feeling for saying the right thing in the right way at the right time. Just let me run through a superficial comparison between journalism and technical writing to show what I mean. In fact, let us limit the comparison to just two points of style—Point of View and Tone.

Let us take Point of View first. Both the tech writer and journalist assume the role of an objective observer. They are supposed to report, not judge or interpret. But you know as well as I do that they both interpret simply by virtue of having to select the details they include in their writing. But the big difference is in the amount of distortion they commit. The news writer is bound to distort more because he is limited by available space and by the interests of his readers and employers. The tech writer, on the other hand, looks upon distortion as the shortest road to professional suicide (one reason, incidentally, why engineers hate to put anything on paper). He is obligated by professional ethics to report *everything*—negative as well as positive results, disadvantages as well as advantages of his product, his mistakes as well as his glories. He must let the facts speak for themselves and the chips fall where they may.

This need for objectivity and thoroughness affects the tone of technical writing adversely. Furthermore, the tech writer's relationship with his reader is on a strictly professional basis whether he is writing reports, letters, or memos. The colloquial, easy-going style of the journalist would be inappropriate, too fuzzy for a professional audience.

This does not mean that technical writing always has to be dull and nothing but dull. Frequently it can be good and dull. But the technical writer is dealing with a complicated subject, unfamiliar to his readers (even when they are as specialized as he is). He must maintain dignity and objectivity appropriate to his profession. He obviously cannot hope to be as entertaining as his journalist brothers, nor as good a salesman as his business-writing brothers. It would be unfair to expect him to compete with them in their own ballparks, playing according to their respective ground rules.

Writing is writing? There are as many different types and forms of writing as there are people in this room. Writing is dynamic—not static—it changes ground rules every time someone puts pencil to paper to compose a sentence. It is taking the easy, and vicious, way out to say that Writing is Writing and then go ahead and judge by arbitrary standards drawn from one type of writing and applied to all.

The man in the street has the ingrained notion that somehow all writing is the same and that Writing Equals Grammar. This is nonsense. Our ivy-covered world is filled with linguistic specialists whose specialty is grammar. They teach what is called the New Grammar, based on solid, scientific linguistic science. They know grammar inside, outside, and backside, but they would be among the first to insist that there is no positive correlation between a knowledge of grammar and the ability to write.

APPROPRIATENESS—THE ESSENTIAL CRITERION OF GOOD WRITING

How, then, does the English teacher judge writing? He judges writing on the basis of whether it has Unity, Emphasis, and Coherence—whether it has an obvious preconceived design, pleasant continuity, and easily understandable language. But he also judges it for appropriateness—and this we ought to discuss at length.

There is a great deal more to writing than understandable words in an understandable sentence. We sometimes forget that behind all the complexities of language is the simple fact that language is a social custom. We are not born with it—we acquire it along with mumps, measles, and falling in love. But we have made an occupational tool out of a social custom, and sometimes forget to carry over little courtesies that make life worth living in a social world. And that is why to this English teacher, at any rate, appropriateness is more important than Unity, Emphasis, Coherence, grammar, punctuation, and any of the mechanical forms imposed by the occupational specialties of writing. Follow the SAP formula: "Writing must be appropriate to S-Subject, A-Audience, P-Purpose." For if writing is appropriate it will be understandable.

This kind of appropriateness is habitual with experienced people, but for the beginner it requires guidance, practice, and sympathy. I was wakened at three this morning by our milkman, wanting to know if my technical-writing-type note meant that we *had* no milk or we *wanted* no milk. The heart of the matter, you see, is really common sense plus writing sense—saying the right thing at the right time in the right way. While most of us have this sense in speaking, how many carry it over into writing?

I am not arguing for a resurgence of that "write like you talk" business. On the contrary, I am one of its most violent opponents because, as a blanket formula, it violates the spirit and the letter of the SAP formula. So does the "folksy" attitude in business letters. Both have been run so deeply into the ground that they now look like ghosts of the past.

Our psychological friends tell us that we cannot tell a man something that is 100% new to him. He must be able to relate whatever we tell him to something he already knows. Then the amount of new information he can take in will depend on the number of possible alternatives he can attach to any given words. For example, if I say: "Shall I stand up or sit . . ." naturally we all fill in the blank with the word "down." We do this because the pattern "sit down" is so common that it becomes a verbal habit.

As you know, the most common verbal habits in English have already been accumulated, sorted, punched-out and fed into IBM machines to serve the Defense Effort through machine translation. One of my graduate students is currently engaged in compiling a glossary of the verbal habits of the machines themselves. Tit for tat. The machines prefer words like "aforlap" instead of "before" and "overlap." They prefer "howby" to "how" and "by"; "timafor" instead of "time" and "before." Obviously, machines that can dream up words like those have a great future—or is it "gruture"? But never fear. I doubt if a machine will ever replace a good writer in producing understandable language. For in addition to requiring understandable words put together in a familiar form or pattern, understandable language also requires what I call "music" behind it.

Tech writing, business writing, journalism all have certain amenities—certain formalities, traditions, customs that make them distinctive. Their words and music are different, but they do have a strong bond in common. They are commercial activities manipulating an awkward social tool. Each uses certain conventions peculiar to itself because these have been found to pay off. Their readers expect these conventions and use them as guideposts. The writer who does not satisfy his readers' expectations will not be a professional writer for long.

As our world unwinds the coils of specialization, more and more we are going to have to broaden the base of writing for different subject matter, audience, and purpose. But meanwhile I think we can live very comfortably with our three-horned dilemma if we do not try to be jacks-of-all-trades, do not expect our tech writers to write like Ann Landers or her sister Abby, do not expect our business writers to write like Hemingway or our journalists to write like Vladimir Nabokov.

DISCUSSION AND ACTIVITY

1. According to Zall, how do technical writing, business writing, and journalism differ from imaginative literature? How do technical writing, business writing, and journalism differ from one another?

2. Explain the SAP formula. Why do writers in business and industry consider the formula essential for effective writing while writers of imaginative literature may not?

Selection Four

Industry Views the Teaching of English

EVERETT C. SMITH

When he wrote the following article, Everett C. Smith was supervisor of employment for the General Electric Company at Fort Wayne, Indiana. Smith shows that the ability to write and speak well enhances your value to businesses and industries.

Let us consider some of the aspects of industry which are perhaps not so well known by those not engaged in it. Let us consider specifically the small manufacturing group which is under the supervision of a foreman. Fifty years ago, the foreman was the complete boss in his group. He obtained the raw materials, he scheduled his production, assigned tasks, inspected the work, kept records of costs, and maintained his tools and machines. Now, however, most of these duties have become so complex as to require specialists. Thus, we have planners who design tools, methods, and layouts. There are production control men who order the material and schedule production. There are wage rate men who classify jobs and set a pattern for placement and payment of the foremen's employees. There are safety engineers, labor relations experts, and cost control men, all of whom have their fingers in the pie. None of these men work directly for any of the others. They must all work together, each trying to fulfill his own particular responsibility, but cooperating with the others.

The confusion could be terrible. But if effective communications are maintained among these different people, if they carefully explain their positions, the concerted efforts of these specialists can be very effective indeed.

Thus the planner's success depends not so much upon his skill in devising new tools and methods, though this is primary and basic, but chiefly upon his ability to sell his ideas to the foreman, the production control man, the cost man. The production control man is not successful because of the accuracy of his records or the ingenuity with which he matches customer's requirements with material availability and machine capacity. Instead he is successful chiefly because of his ability to expedite, to convince the others on the team to follow his schedules and accept his manufacturing priorities. So it is with the safety engineer, the wage rate man, and the labor relations man. Their success depends not so much upon their special knowledge or technical skill, but upon their ability to present, to sell their advice.

In order that this be not entirely hypothetical, I took a little survey among several of my associates. I asked them what percentage of their working time was spent in the following: reading, writing, listening, talking—and all others. I asked an engineer, an education and training specialist, a superintendent, a labor relations negotiator, and a cost leader. The average came out this way:

Reading	22%
Talking	21%
Listening	15%
Writing	13%
All Other Activities	29%

Obviously, there is a considerable amount of thinking, judgment, and analysis going on in conjunction with the reading, writing, listening, and talking. The significant thing for us, however, is that these men are using language as their tool in seventy-one percent of their activities. Just as the mechanic must be skillful in the use of his wrenches and screwdrivers, so these men must be skillful in the use of words. Now, granting that men working in industry must be able to present their ideas effectively, does it necessarily follow that they must study English in order to develop this ability? Do students think so? They might well ask:

Sure, I know that a man has to be able to get his ideas across, but does he need to read Shakespeare's sonnets or learn to conjugate verbs in order to express himself? I talk with my family and friends every day and I get along all right. All you have to do is say what you mean! That's not hard. I don't have to study English to do that!

Let me read you a sentence from one of our business letters. This was written by an experienced, capable man in our sales organization, who knows his job and knows what he is talking about. He is discussing the change-over from a #6 to a #8 cable on one of our motors.

If Wright Field had had reasons for the use of #6 due to the fact that the motor might be used most severely at times since if the C26 ground power unit had become mared [sic] or for any such reason was hard to move you can bet your life that the sergeant who was ordering it moved over to the next engine would not be concerned about the overloading of the motor, but would expect the unit to move if the voltage was cut down by this motor lead so that the motor could not get enough amperes to develop [sic] sufficient torque to move it out of the hole it was in, then it would be a lot of trouble starting with the pilot in the plane who wanted to get in the air and shoot down somebody who was coming after him.

Yes, he knows what he is talking about, but does anybody else know? Did he find it easy to "say what one means"?

At one stage of my industrial experience, I was a raw material leader. I had three men working under my direction. Our job was to determine the factory's needs for materials and place orders on approved vendors for these materials, specifying exact dates of shipment. Then we had to follow up to make certain the material would be shipped as needed.

One of these men had been with the company for twenty-five years or more. He had done the same job for a long time, he knew the routines, he was good at figures, and he understood our objectives. He did an adequate, though uninspired, job. One day he called to my attention an invoice which we had received from a vendor, billing us for some material we had ordered. The material, which normally arrived about the same time as the invoice, had not come in. So we could not approve the invoice for payment. Yet if we did not approve it and send it along, we would likely lose the discount afforded for prompt payment. Already we had waited four days in the expectation that the material would arrive.

I told Bill to write a letter to the vendor. He was to ask if shipment had been made and if so, what the routing and way bill were. Then we could trace the shipment. If shipment had not been made, when would it be, and should we return the invoice for re-issue at the time of shipment? This is what he wrote:

Refer our order No. FR xxx. Your invoice IL000 received. When will you ship? Urgently required.

Very truly yours,

Bill knew the situation. He was the one who had brought it to my attention. Yet he could not explain it. Fortunately, he showed me the letter before it went out, and we sent another.

LANGUAGE TRAINING NEEDED

Last year in our company all of our supervisory personnel had to fill out position descriptions. These were thirteen-page questionnaires concerning our jobs. The purpose of them was to evaluate our jobs and so put them in the proper place in the salary scale. Obviously it was to our personal advantage to complete these as fully as possible so as to make our jobs appear just as important and responsible as could honestly be done. A few of the questions could be answered by checking blocks, but the majority required paragraphs of explanation. For most of our people it was a grueling task. One would think that surely a man could write about what he did every day. There are few things with which a person is more familiar. Yet for those who had not had the benefit of English instruction, it was one of the hardest jobs they had ever done. It was not really easy for the rest of us.

No, the evidence is overwhelming that it is not easy to say what one means. Facility in self-expression does not come naturally.

Peter Drucker, in the May 1952 issue of *Fortune*, had an article on being an employee. A particular point that caught my fancy was his conclusion that poetry writing and short story writing were probably the most important subjects taught in our schools from the standpoint of preparing a person to be a good employee. Mr. Drucker went on to say that an employment manager would likely be highly unimpressed by a young applicant who presented as his chief qualification a college major in poetry writing and short story writing. Yet that same manager would be the first to complain that not a person in his organization could write a decent report. These subjects, Mr. Drucker says, develop a sensitivity to language. Sensitivity to language— perhaps that is the key to the whole problem.

So much of our trouble with other people comes from lack of sensitivity to language. We fail to recognize the connotation which certain words have for other people. As a result we may antagonize them or prejudice them against giving us their cooperation. For instance, The Opinion Research Corporation published in its Public Opinion Index for June 1954 some interesting answers to differently worded questions. Ten hundred and seventy-one working people were polled: salaried, hourly-rated, and piece workers. To the question, "Would you like your company to move in the direction of more improved machines that do the work quicker and better?" sixty-two percent answered "Yes." To the question, "Where companies put in improved machines that do the work quicker and better, do you think it is a good thing for employees?" fifty-two percent answered "Yes." Yet when the question was asked, "Would you like your company to move in the direction of more *automation*?" only thirty-one percent answered "Yes."

There are many such words around which attitudes have developed. We cannot effectively lead, guide, and explain unless we are sensitive to words, to implications as well as definitions.

If you accept my evidence as so far given and my conclusions therefrom, you might still ask: "Is industry, however, sufficiently convinced of this to do anything about it? Are business leaders taking any positive steps to improve their own English and that of their subordinates?" That, of course, would be the real proof of our sincerity, of our conviction.

The answer is yes. As evidence, there is Toastmasters International, formed by businessmen as a means of improving their ability at self-expression. Also the Forum Club has as one of its objectives the development of its members' ability to speak effectively.

In my own company we have several formal courses for selected employees designed to improve their ability to write and talk effectively. One of them is called "Effective Presentation," a course which runs for seventeen weeks, meeting once per week for two hours. All class time is outside of regular working hours, that is, on the employees' own time. It was started in 1943 in an eastern plant by a group of sales personnel, who felt the need of some means of developing their ability to express their ideas. In eleven years the course has spread to all major works of the company, with over a hundred separate classes now going on. There are over twelve thousand "graduates" of the course.

The significant thing about this rapid growth is that, although the company management has fostered the course, the chief impetus has come from employees themselves. Although enrollment is still limited to supervisory, specialist, and trainee personnel, there are more candidates each year than can be accommodated.

For the course we have a specially prepared text in which we give reading assignments. There are several assignments in letter writing, such as:

1. Write a letter to a business associate commending him on a good job well done.

2. Write a letter to an employee in your unit, who is retiring from the company.

3. Write a letter suggesting some innovation in the procedures, policies, or working conditions in your unit.

The most important part of the assignment, however, is the preparation of a three-minute talk by each member of each class. A written outline and the written talk are both required to be turned in. These are conscientiously corrected by the instructor and returned. The talk itself is delivered to the class, after which the class and the instructor criticize the speaker.

It is a rather rigorous course. Yet it is the most popular course that our company offers. One of the men in my class this year, who had finished high school some twenty years ago without having paid much attention to English or any other subject, told me that he had been pestering his superintendent for two years to get him into the course.

Yes, both management and employees recognize the need for improving their ability in expressing ideas.

INDUSTRY AND THE ENGLISH CURRICULUM

I suppose it is only human nature that I should wish to take this opportunity to deliver myself of a few comments on some phases of the English curriculum such . . . as grammar, composition, literature, public speaking, perhaps poetry writing, journalism, and short story writing.

Literature might be presumed to be quite irrelevant to an individual's success in industry. Yet I believe it has its place in the English curriculum, not just for its cultural value but also for its practical value. I honestly believe that one cannot read Ruskin or Emerson, Wordsworth or Whittier, however cursorily, without some of that elegance and facility rubbing off. It is true that we learn mostly by practice, but we learn also by precept. Good literature affords the precepts.

Public speaking might be the most useful course in the English curriculum, so far as industry is concerned. I do not know how it is taught now, but I recall how it was taught when I went to school. I took one semester of public speaking in high school, and I gave four talks in the entire semester. The rest of the time I listened to or slept through talks by classmates or instructions by the teacher. If that is the way public speaking is taught now, I doubt its efficacy. I believe a student should give a talk at least twice a week. This is one subject where we need a maximum of practice and a minimum of precept.

On poetry writing and short story writing, I will rest my case with Mr. Drucker. If these subjects develop sensitivity to language, then let us have them.

In grammar, there is much that is trivial. I fail to see how clarity is lost if one says, "It don't" instead of "It doesn't." Nor does it seem really to matter whether I say, "Everybody will take their own notes" or "Everybody will take his own notes." I suppose this sort of thing is more a matter of etiquette than of efficiency, like eating peas with one's knife. But when we deal with sentence construction, grammar becomes very important. So many sentences that I hear and read in industrial correspondence are so involved and grammatically incorrect that it becomes an exercise in logic to decode them. I like to think of grammar in connection with the little coin changers worn by our city bus drivers. Imagine, if you will, what happens when the bus driver receives his change in the morning. If the cashier hands him a roll of half-dollars, a roll of quarters, a roll of dimes, and a roll of nickels, the driver's task is simple. He merely removes the paper and puts each stack of coins in the appropriate slot. It is quick and does not even require his concentration. But if the cashier handed him a bag of loose coins, then the driver would have to sort them out individually and put them into the coin changer. It would take eight or ten times as long and would require his close attention.

So it is with a sentence. If we hear a well-constructed, grammatical sentence, the ideas fall easily and quickly into the slots of our consciousness. But if we hear a conglomerate, ungrammatical hodge-podge, we have to sort it out at an expenditure of time and effort. By the time we have done so, the speaker is three or four sentences beyond us, and we have lost the thread. After only a few such sorting exercises on the speaker's sentences, the average listener decides the effort is not worth it, and goes to day-dreaming of what he will do when he gets out of the meeting.

. . . Perhaps the best reason for studying grammar is summed up in the following statement: "What I have to say is so important that I want it expressed in the best possible language."

Composition is the course in the English curriculum which we most favor. It is the very fabric of what our employees are using every day. There is where the student should apply himself most vigorously and where the teacher should instruct and motivate most effectively. . . .

In industry when Salesman Jones writes a letter to Foreman Brown, pointing out certain deficiencies in deliveries and suggesting remedial procedures, he is committing himself as to:

1. The accuracy of his information.

2. A course of action which, if adopted and proven ill-advised, will reflect to Jones' discredit.

Therefore, Jones will be careful to say exactly what he means in terms which cannot be misunderstood or misrepresented. It is this feature of going out on a limb every time one puts something in writing that characterizes business correspondence. Some people react to this form of putting the finger on one's self by avoiding all written communication, thereby greatly curtailing their effectiveness. Others, recognizing the need for a positive approach to their jobs and responsibilities, find a very strong motivation for improving their ability in self-expression.

. . .

English, as a subject, seems to have waned in popularity during the last generation. Most students . . . seem to regard it as something that is taught because it has always been taught, that it has no real practical application. It would be much better, they think, to study shorthand, typing, machine shop, science and mathematics, because those subjects prepare one for specific jobs. Industry has contributed to this attitude. Even in times of curtailed production, we have still hired a few stenographers, engineers, accountants, and toolmakers. Whereas, even when we are hiring heavily we have very few jobs for the . . . English [major].

Yet this is not a weakness in the case for English. Indeed, it is quite the opposite. English is not something used by a few specialists but by everyone: the stenographer, the engineer, the toolmaker, the accountant, the clerk. Ability in English enhances one's value in all these fields. It is probably not the wisest course for a student seeking a career in industry to major in English. It is better to select a specialty, but to take English as a supplement. My point is that English is a supplement needed and used by all specialists.

The best toolmaker will never become a successful foreman purely on his technical ability. The most proficient engineer will not make a supervisor purely on his engineering know-how. Something else is needed. We call it "leadership." There have been a lot of attempts to define leadership, some adequate, some not, none complete. One ingredient which seems common to most is the ability to express ideas to others. I know of no better way to develop this ability than by the conscientious study of English.

DISCUSSION AND ACTIVITY

1. What motive does an executive from industry have for writing an article encouraging students to develop a mastery of English?

2. Reread the sentence from the letter discussing the change from #6 to #8 cable (p. 24). Revise the sentence to make its meaning clear.

Selection Five

English Skills Among Technicians in Industry

HAROLD P. ERICKSON

Harold P. Erickson teaches English at Coleman Technical Institute, La Crosse, Wisconsin. According to his report here on a questionnaire sent to almost four hundred industries, the ability to read, write, and listen well plays a significant role in the work of a technician.

In the rapid growth of the technical, vocational, and adult schools in Wisconsin, the technical subject areas have kept pace with the constantly changing demands of industry while some of the academic areas, such as English, have had little or no revision to meet these new demands. For this reason, 379 questionnaires were sent to industries in Wisconsin, Iowa, and Minnesota to determine the English needs of the technician. In addition, fourteen interviews were conducted with businesses in Wisconsin and Minnesota. The results of this study are recorded in Tables 5-1, 5-2, and 5-3.

The responses indicate that between 57.2 and 69 percent of the technician's time is devoted to the use of communicative skills. They also suggest that technical reporting represents a major portion of the technician's job, and that this task is about equally divided between oral and written reporting.

BUSINESS LETTERS

Table 5-1 shows that 64 percent of the responses indicated a mean average of 6 percent of the technician's time is devoted to business correspondence. Further discussion with the companies revealed that these business letters generally fit into three categories. First, he must be adept in composition of letters of inquiry, biddings, and requests for specifications. Since the techni-

Table 5-1   Results of surveys

| | RESPONSES FROM INDUSTRIAL QUESTIONNAIRES | | | INTERVIEWS |
TASK	NUMBER OF RESPONSES	% OF USABLE RESPONSES	MEAN % OF TIME	MEAN % OF TIME
Business letters	86	64	6.0	10.0
Writing technical articles	53	39	11.0	00.0
Reading technical articles	129	96.9	10.1	16.7
Oral communication	127	95.4	30.1	42.3
Total time spent in communicative tasks			57.2	69.0
Total time spent in technical tasks			42.8	31.0
			100.0	100.0

Table 5-2   Technicians in research work

TASK	RESPONSES	% OF USABLE RESPONSES	MEAN % OF TIME
Written reporting	113	84.9	24.5
Oral reporting	93	69.9	25.4
Communicative tasks			49.9
Technical tasks			50.1
			100.0

Table 5-3   Desirable English skills

SKILLS	% OF RESPONSES
Spelling	69.8
Technical report format	86.0
Reading	65.0
Vocabulary	75.4

Note: These three tables represent the results of 133 returned surveys and 14 ninety-minute interviews with companies in Wisconsin, Iowa, and Minnesota. Each company had to employ technicians in one or more of the following fields: automotive, electronics, mechanical design, air conditioning and refrigeration, and/or commercial art.

cian is involved in development and maintenance, he must be able to communicate with interplant personnel as well as with other companies to inquire about new equipment. Secondly, the technician who works for a small company must often double as the purchaser of testing equipment or replacement parts. The letter of purchase, with its exactness and conciseness, should be included in any training program for the technician. Finally, the technician should be trained in the correct use of application letters. All of those personally interviewed and several of the written questionnaires indicated that few technical school graduates could write an acceptable letter of application.

In addition to letter writing, the personal interviews indicated that the technician spends a fair amount of time in writing interoffice memorandums, which can also be classified in the business letter category. Since the questionnaire asked only about business letters, the omission of memos could be a reason for the rather low percentage of time assigned to letter writing in the responses.

WRITING OF TECHNICAL ARTICLES

39.0 percent of the responses indicated a mean average of 11 percent of the technician's time is given to the writing of technical articles for publication. On the basis of this response, it appears that this task has little significance for technicians among those industries surveyed. One reason for this may be that technical articles are generally written on topics of product research and development; therefore, this task is usually left to the engineer. Those companies that responded affirmatively had the majority of their technicians working in research and development sections.

Only one interview gave any support to the need for writing technical articles. A company that employs a staff of eighteen commercial artists and draftsmen in its art department gives them the responsibility for the editing of the company paper on technical matters. The technicians have to be able to gather information and condense it for a journalistic report. Also, the commercial artist or draftsman is involved in the preparation of sales catalogs, which requires some technical descriptive writing.

READING TECHNICAL ARTICLES

96.9 percent of the responses submitted reading as a vital skill for every technician. Most of the interviewees suggested an average of between 10 and 20 percent of technician time expended in reading, while the survey showed a mean average of 10.1 percent. This average reading time on the job seems rather low considering the numerous responses and as the importance of reading was stressed both in the comments and the personal interviews. Many of the firms that were interviewed indicated that the technicians were encouraged to read trade journals and product releases for specific job-related information. Also, due to the high complexity of technical vocations, there are unlimited numbers of specifications, instruction manuals, and company directives to follow and understand.

Comprehension was the key word in reading skill, rather than speed. There can be no doubt that the technician is severely handicapped if he lacks the ability to read for accuracy and understanding. Evidence of the importance of comprehension in the reading of technical material was indicated by 65 percent of the respondents who desired further training in reading for the technical graduate.

ORAL COMMUNICATION

95.4 percent of the companies surveyed indicated that a mean average of 30.1 percent of the technician's time is spent in oral communication. The personal interviews suggested that this estimate might be low by as much as 15 to 20 percent. During the ninety-minute interviews, it was revealed that more than 30 percent of the technician's time is taken up in oral communication.

Since the technician provides liaison between the production worker and engineer, a good portion of his time is spent in either sending or receiving oral directives. If the technician happens to be working in a mechanical design section, it is not unusual to see him discussing plans with engineers for fifteen to twenty minutes at a time. Technicians working in the plant as testing personnel keep an almost continual technical dialogue going as they report test results. Those companies that use technicians as servicemen in the various utility fields indicate that their technicians must be able to meet the public and represent the company with courteous service while being able to explain the mechanical aspects of a problem in laymen's terms.

Some rather general oral tasks cited by the respondents are the conducting of product demonstrations, sales meetings, and safety programs. Seven companies had their technicians actually staff exhibition booths at county fairs and industrial shows. Again, the ability to speak to groups of people is considered very important.

The skill of effective listening, although not listed on the questionnaire, received enough comment in both the written responses and the interviews to be mentioned in this report. Perhaps the best summary of the feelings of industry regarding this skill was explained by a respondent who concluded: "The listening aspect is most important. Communication, to be highly effective, must totally encircle sender and receiver." In the work of the technician, where thousands of dollars may be invested in each project, the ability to listen to and comprehend instructions plays a major role.

TECHNICAL REPORTING

Both the comments and the interviews support the conclusion that a total of 49.9 percent of research work is spent in reporting. This is further broken into written reporting, 24.5 percent, and oral reporting, 25.4 percent. Formal or informal technical written reporting generally falls into two classifications: description and analysis. In the descriptive classification, the technician may do several possible types of work. Mr. Klaus, chief engineer, La Crosse Telephone Company, stated that reporting ability was a major factor in his company. Those technicians who perform service work away from the plant must report every line breakdown that occurs, with explicit instructions on the repair of the problem or recommendations for the next shift. In the case of the night shift, where there is often no engineer on hand, the importance of the report is increased. In most cases, this report is made on a standard form. At the end of the month the technician prepares a formal summary report ranging from one to ten pages and in which clarity, conciseness, and accuracy of language are essential.

Changes in product specifications are another source of descriptive report writing for the technician. In a large drafting and development section, each print can have as much as 10 percent of its area devoted to instructions. The ability to accurately instruct the user of the print becomes as important as the print itself.

Finally, the descriptive type of reporting is present in the writing of sales brochures. A smaller company almost exclusively used its technicians to write not only the descriptive literature of the product but also the instruction sheets and specifications for the customer. For larger products, the report may run a total of eighteen to thirty pages, while the smaller items can be easily handled in one page. Whatever the case may be, the emphasis must be on the ultimate use of the material by the consumer.

The second major category of written reports is the reporting of test analysis. This is generally done on printed report forms on which the technician is responsible for accurately reporting data he receives from the tests. Some of the larger companies that were interviewed made extensive use of computers to handle detailed work, but there were still many instances where physical tabulating and reporting were being done by the technician.

Written reporting includes not only formal research reports but also the daily reports that the specialist must perform. The amount of time spent in reporting these tasks is directly related to the role of the technician. For example, technicians employed in the research and development areas, who spend the majority of their time in pure research, have a definite need for technical writing abilities.

Another factor in report writing is the necessity for accurate spelling. The survey showed that 69.8 percent of the companies indicated a need for spelling improvement among technicians. Every interview supported this observation, and comments on the questionnaires stated that the technician's spelling ranged from "poor" to "atrocious." Personal observations showed the undesirable effect a misspelled word could have on a blueprint that is sent to a customer.

An unnamed manager summed up the entire philosophy of the written report in these words: *An individual's work is . . . questioned if his work is sloppy, made up of obvious errors, and contains spelling errors of simple, common words. If there are errors in the written presentation, there is an inference of errors in the original work.*"

ORAL REPORTING

Approximately 70 percent of the companies indicated a mean average of 25.4 percent of the technician's time is given to oral reporting. The interviews definitely showed oral reporting to be a major factor in research work. It is nearly impossible to separate conversational reporting from factual reporting to a superior or subordinate. However, one major point to consider is the level of appropriateness in the language the technician encounters. Even though some of the oral reporting tasks he may perform are of an informal nature, he nevertheless must be able to comprehend and speak in the vernacular of all levels, from the engineer and the administrator to the worker on the line. There is further evidence in the comments on the questionnaires indicating that many industries have their own terminology; therefore, the technical graduate must be able to learn new terms peculiar to his field. Stronger preparation in the skills of basic and specialized vocabulary building were indicated in 75.4 percent of the industries surveyed.

Whereas the written report is a permanent record of the technical research being performed, the oral report is an on-the-spot report of progress being made. According to the supervisors interviewed, it is a common occurrence for a technician to be called upon to explain the progress of an experimental or developmental project before a directors' or managers' meeting. The need for training in short impromptu speeches is very apparent in this function of the semiprofessional worker.

More formal oral reporting involves the effective use of graphic materials and some training that will aid in their use and preparation is recommended. Once again, this differs with various fields, but on the whole, the technician should be able to effectively illustrate his presentation with whatever visual or communicative aids he has at his disposal.

The oral report fits the same categories as the written; that is, reports of description and analysis. The major difference is one of preparation and delivery. While the written report can be prepared, reviewed, and then rewritten, the impromptu report must be logically organized and clearly delivered on the initial attempt. Furthermore, there are distinct differences in kinds of language, color, and effect between oral and written reports.

DISCUSSION AND ACTIVITY

1. Although technicians might expect to spend most of their time at work exercising their technical skills, according to Erickson's report they will spend between 57.2 and 69 percent of their time using their communicative skills. In view of such findings, is the mastery of a technical specialty alone sufficient for making a technician a valuable employee?

2. Ask employers in your area of specialization how much time their employees spend writing letters and technical articles, reading technical articles, and communicating orally. Compare the figures with Erickson's.

Selection Six

What Price Verbal Incompetence?

EDWIN A. LOCKE, JR.

> *When he wrote the following article, Edwin A. Locke, Jr., was president of the Union Tank Car Company. His article cites several instances in which verbal incompetence proves damaging and costly to individuals, businesses, and even countries.*

It seems to me that I have never heard so much misuse and abuse of the English language by people occupying responsible positions as in recent years. And my question is: What ought we to do about verbal incompetence—about the carelessness in speaking and writing which is so prevalent among us?

It is not exaggerating to say that America's international position depends in some degree on our ability as a people to use words effectively and grasp their actual meanings. We have entered a period of greatly increased competition from other countries. Our qualifications as leader of the free world are being tested as never before. Our diplomats are constantly in the spotlight. Every word that they utter or write is exposed to hard critical judgment. Our government must compete in propaganda with other nations to influence the world's peoples, and our success or failure in this department can profoundly affect our destiny. If we are to maintain leadership, we will have to show great skill in the arts of communication. Those who do not understand us may well turn against us.

I shall never forget one incident of the days when I was working for our government in a diplomatic capacity. There was a meeting in London where a very high official of our government had to explain our position on a touchy matter to the ministers of several other countries. I think all of the Americans

From *Harvard Alumni Bulletin*, February 1962. Copyright © 1962, *Harvard Alumni Bulletin*. Reprinted by permission.

present, including myself, were embarrassed for him and for the United States. He rambled, he stumbled, he used words that distorted his meaning. He not only failed to make his point but he confused the issue. It literally took weeks of patient effort to correct the false impressions that he created at that meeting, and we never did win the agreement that we sought.

I have heard a Congressman shock a group of Asiatic visitors by saying proudly that Americans are an "aggressive" people. I suppose he meant to say that we are a courageous, vigorous people. But his audience understood him to mean that we are a warlike people, and their worst fears were confirmed. I hate to think of how many times in our recent history similar episodes must have occurred—meetings where our spokesmen used the wrong words, or failed to use the right ones that would have made their meaning clear. These are days when, as I read the newspapers, I find myself repeating the closing lines of Kipling's *Recessional:*

For frantic boast and foolish word
Have mercy on thy people, Lord!

We are told, and I think we can all agree, that verbal incompetence is often a symptom of deeper problems of the mind and spirit. But there are surely many instances when it results from loose and undisciplined mental habits, from a lack of respect for words. Unless we take action to correct the trend, verbal incompetence may become a national calamity.

Lately I have come to think that the sector of our society that suffers most seriously from verbal incompetence is business. There, too, the national interest is involved. The misleading use of words is a major source of inefficiency and waste motion in business. It is all the more serious because it is hard to measure. I venture to say that all the thefts and embezzlements that corporations suffer every year do not cost our economy as much as verbal incompetence.

The waste created by misunderstanding is one that America can ill afford. Our industry is now compelled to meet concentrated competition, not only from the Communist countries, but even more important, from nations friendly to us. High prices—and waste means high prices—can keep us out of markets all over the world. And if our foreign trade falls away—or even fails to grow—our political influence and strategic alliances could be jeopardized.

All of us, in whatever walks of life, have a big stake in the efficiency of American business. And for that reason, we must take a serious view of the continuous economic losses due to verbal incompetence.

I have seen more than one business shaken by a single letter or memorandum in which words were used loosely or wildly. A story was recently told to me by the head of an important company. Call him Mr. Brown. He was at the time trying to establish friendly relations with an executive of another company, a Mr. Slade, who was an important potential customer for Brown, and Brown had given a good deal of thought to the best way of cultivating him.

One day, a letter from Slade arrived at Brown's office. Slade said that he was reviewing his requirements for the year ahead, and if Brown would like to talk with him, he would make himself available.

Now it happened that Brown just then was away on a trip. In his absence, Slade's letter went to one of his young assistants for reply—call him Harvey. This is what Harvey wrote to Slade. "Dear Mr. Slade: In Mr. Brown's absence, I am writing to say that your request for an appointment will be brought to his attention immediately on his return."

When Brown got back a few days later, he telephoned Slade at once. Slade shocked him by saying he was no longer interested in pursuing the matter. He said that he judged companies by the tone of their correspondence, and after receiving Harvey's note he had got in touch with another company, a competitor of Brown's, and had concluded a deal with them. He added that he had been surprised to find that his letter to Brown was regarded as a request for an appointment.

When Brown hung up the phone, he sent for Harvey. Now the point that interests me most is that Harvey could not see that there was anything wrong with the letter he had written. He said, "But Mr. Brown, that letter from Mr. Slade *did* ask for an appointment."

Brown said, "You don't seem to understand. Slade wrote that if I wanted to see him, he would make himself available. He wasn't requesting an appointment. He was giving *me* a chance to request one."

And then young Harvey said, "But after all, it's practically the same thing, isn't it?"

Business is full of Harveys—young men whose minds have never been trained to pin-point the meaning of the words they use, and whose careers suffer accordingly. That little anecdote is characteristic of what goes on all the time. A man can have many virtues and abilities, but if he does not use language accurately and carefully, he can be a positive menace to a business enterprise.

This is a country where each year hundreds of thousands of young men go into business with their hearts set on executive careers. Yet it is relatively rare to find a young businessman who recognizes how much his chances for an executive post depend on his ability to use words effectively. Many companies, including the one with which I am connected, today give increasing weight to the ability to communicate effectively as a test of executive potential.

Of course, there are plenty of young people who have a bright surface and glib tongues—who look at first glance as if they might be of executive caliber. But when they begin to talk seriously, or to express themselves in writing, they too often reveal serious verbal limitations. Sometimes, listening to a hopeful young man, or reading something he has written, I have been reminded of the way a child uses finger paint—a dab here and a smear there. The child hopes that Father will recognize his painting and say, "Oh, yes, that's a cat." Many otherwise intelligent people, when they talk or write, similarly seem to feel that they have done all that can be expected of them if the other fellow just gets the general idea. They may not know it, but they are intellectually crippled.

The heads of several other companies have told me that they, too, are deeply concerned over this problem. They have all had experience with young executives who seem almost indifferent to the meaning of words. I have been told of a memorandum written by one such young man, which caused a good deal of trouble. His company specializes in issuing reports on the steel industry. One day the editor of these reports received a memo from one of his assistants. It said, "I talked with Anderson, and he predicts a price rise before April." Anderson was an important steel man.

The report for that week was going to press, the editor was in a hurry, and he did not stop to check with Anderson. In a few days the item appeared in print, and then Anderson telephoned in a state of rage. He demanded an apology and a retraction. He had made no prediction, he said. All he had told the young man was that if certain things happened—which had not yet happened—prices could be expected to rise. Since when was that a prediction? If that was the kind of intelligence the editor was relying on, said Anderson, he could never trust him again.

Now the young assistant who had written that memo knew in general what the word "predict" means. But he did not distinguish between a prediction and a qualified statement of possibilities. He was content to use a word that merely approximated his meaning. As a result, he did serious damage to the reputation of his firm.

If verbal incompetence were confined to the use of the wrong word, it would be bad enough. But equally dangerous is the frequent inability of businessmen to sense the effect of their words on the persons who will hear them or read them.

I have known a single word, used insensitively, to touch off a costly labor dispute. A large company was negotiating a new contract with a labor union. Both sides had presented their views in writing. The negotiation was progressing in a somewhat tense but reasonable atmosphere. At this point, the union leadership presented a letter raising a new condition. The letter hinted that if the new condition was not accepted by the company a strike might result. This was, of course, a familiar bargaining tactic.

The union's letter was given for reply to a young man in the company. By current standards, he is well educated—a graduate of a great university— with a good academic record, and a lively mind. The letter he composed was for the most part sensible. But at one point he said, "It would be criminal to call a strike for such a reason." None of his superiors who read the letter saw anything wrong with it, and it went out.

Now it was true that a strike under those circumstances would have been illegal. The young man knew the difference between "illegal" and "criminal." To do him justice, he used the word "criminal" only in its figurative sense, to show indignation. Unfortunately, the union leaders took it literally. The word "criminal" was like a slap in the face to them. One of them, in fact, had a jail record. They reacted violently to what they felt was a gratuitous insult. "Who is he calling criminals?" was their reaction. From that point on, they became hostile, the situation deteriorated, and a useless strike followed at great cost both to the union members and to the company.

Another costly aspect of verbal incompetence in business is what might be called the careless cliché. Please understand that I am not objecting too much to clichés as a whole. They may be despised by poets, but as we all know a cliché accurately used can on occasion be a time saver and a boon to the weary mind. The trouble arises when the cliché is wrongly used.

I know about a letter written by the sales manager of a well-known company to a customer. The letter explained why a certain salesman had left the company. The sales manager was angry at the salesman for quitting and in his letter he said, "It's just a case of a rat leaving the ship."

He failed to remember that it is the sinking ship that rats desert. But this thought came to his customer, and he mentioned it to others. Soon people were gossiping that the company was in trouble. It took an investigation to unearth the source of the rumor, and a good deal of effort to undo the damage that had been done. The head of that company told me that he figured the cost of that one little misused cliché at about $10,000.

Then there was a memorandum issued by an officer in another company, with this apparently innocent sentence in it: "Let's apply this new credit policy with discretion right across the board."

The writer of that memo meant that the new policy should be applied in all appropriate cases, but that there would be some cases in which it was not applicable, and care should be taken to handle such cases discreetly. That was not the impression that was created in the minds of the men who received the memo. As they understood the order, the new credit policy was to be applied to all of the company's customers, "right across the board," and the words "with discretion" merely meant that they were to be polite about it.

It took just two days for the earthquake to develop. Then complaints began to come in from outraged customers, demanding to know what the company meant by refusing to extend the usual credit terms. Tempers were lost. Orders were canceled. The company's top management became alarmed. Before the tangle was straightened out, I was told, the company had lost $50,000 worth of business.

To me, one of the most irritating forms of verbal incompetence is wordiness. It is of course hard to be concise. I have a good deal of sympathy with the man who said, "If I'd had more time, I would have written you a shorter letter." We all know what he meant. But a great deal of unnecessary verbiage in business, I am convinced, results not so much from lack of time as from mental laziness or confusion.

More than once I have seen executives spend valuable hours and brain energy trying to grasp the meaning of a ten-page report, when a single page of accurate writing would have served the purpose. I have seen the employees of a plant bewildered and disturbed by a long, incomprehensible instruction from the home office, until it was reduced to a few brief sentences that told them clearly what to do.

Business does not want wordy men, but it urgently needs men who respect words. I hate to suggest that our overburdened schools take on more responsibilities than they already have, yet I cannot help wondering if this is not essentially a problem of education.

People are most responsive to the discipline of language when they are young. In several instances I have tried to get mature business associates to sharpen their use of words, but the results have been a good deal less than spectacular. Once a young man has left school, if he has not already been imbued with respect for words and given the mental training necessary to distinguish shades of meaning, the chances are that he will always be weak in this department.

People who cannot use words accurately are likely to be people who cannot grasp meanings successfully. And much may depend on the ability of the American people to understand the actual meanings of the propagandistic words that are being hurled at them today. Why are so many people everywhere carried away by demagogues and fanatics? In part, I think it is because they have never been exposed to the least semantic discipline. They do not really understand the words that they hear or read. It is only the emotional overtones that reach them. And this unthinking emotionalism can in the long run make them dangerous to their countrymen, to themselves, and to the world.

DISCUSSION AND ACTIVITY

1. Several of Locke's examples of verbal incompetence portray speakers or writers who mean one thing by what they say, but whose audience construes the meaning differently from what was intended. Who has the responsibility to clarify meanings, the speaker/writer or the audience?

2. If you were the head of a company, how much emphasis would you place on your employees' communication abilities as you considered the employees for raises or promotions? Would you promote Harvey to a top executive position?

Suggestions for Further Reading

WHAT IS BUSINESS AND TECHNICAL WRITING?

Britton, W. Earl. "The Trouble with Technical Writing Is Freshman English." *Journal of Technical Writing and Communication* 4 (Spring 1974), 127–131.

Technical writing often differs from ordinary writing in that (1) its emphasis is on readers rather than writers, (2) its information flows from knowledge to ignorance, (3) its assignments are intellectually taxing, (4) its success and effectiveness are measurable, and (5) its rhetorical devices (such as focus, logical partitioning and classification, and illuminating sequence) can more likely be transferred to ordinary writing than vice versa.

Gould, Jay R., and Wayne A. Losano. "Technical Writing—An Important Profession." *Opportunities in Technical Writing Today*, rev. ed. Vocational Guidance Manuals, Louisville, 1975.

"Technical writing . . . is the profession of writing, editing, and preparing publications in many fields of technology, science, and engineering, including technical and scientific magazines. The publications may be *technical reports, instruction manuals, articles, papers, proposals, brochures* and *booklets,* and even the preparation of *speeches* for technical meetings and conferences."

Hays, Robert. "What Is Technical Writing?" *Word Study* (April 1961), 1–4.

Technical writing differs from other prose in two basic ways: one can be called psychological and the other linguistic. The psychological difference is in the technical writer's attitude of "utter seriousness . . . of objectivity (if humanly possible), respect for data and their limitations, and caution." But the greatest difference is linguistic, because technical writing requires a specialized vocabulary, especially of adjectives and nouns.

Robbins, Jan C. "Social Functions of Scientific Communication." *IEEE Transactions on Professional Communication* 16 (1963), 131–135, 181.

Scientific communication, or the broadcasting of research results, has social as well as professional functions. For, besides being a means of extending professional knowledge, communication through writing and speaking is also a means of determining one's professional status.

Weber, Max. "Technical Writing and Human Engineering—An Analogy." *Field Engineers' Electronics Digest* 2 (1963), 31–34.

Human engineering is "the modification or design of something to suit the capabilities and limitations of the human operator." Writers in business and industry can benefit from a knowledge of human engineering, because successful communication is basically a matter of suiting unfamiliar subject matter to the capabilities and limitations of readers.

WHY STUDY BUSINESS AND TECHNICAL WRITING?

Chapman, Gilbert W. "Specific Needs for Leadership in Management." In *Toward the Liberally Educated Executive*, edited by Robert A. Goldwin and Charles A. Nelson. Mentor Books, New York, 1960.

"Management is deeply involved in the art of communication and often success and profit depend upon it. Eventually, all communication must be communicated, either orally or in writing. The ability to express oneself and the ability to understand what is expressed are absolute prerequisites for successful executive performance."

Drucker, Peter F. "How to Be an Employee." *Fortune* (May 1952), 126–127, 168, 170, 172, 174.

The one basic skill to help one in being an employee is the ability to organize and express ideas in writing and in speaking.

Estrin, Herman A. "Engineering Alumni Advice to Freshmen on Studying English." *College English* 21 (November 1959), 98–99.

According to two hundred engineering alumni who were questioned, the ability to communicate more than any other quality that one has determines one's professional achievement. One should concentrate particularly on acquiring the ability to write clear, concise letters and articles. One should learn not only grammar but also rhetoric.

Pearce, Frank. "Desirable Writing Skills for Personnel in Business and Industry." *Teaching English in the Two-Year College* 3 (Fall 1976), 29–31.

Whatever a person's job, he or she must be a master of writing skills in order to advance. One must be able to: (1) identify audience and purpose, (2) organize, (3) be concise, and (4) use plain language in a straightforward manner.

Reynolds, Neil B. "Why Study English?" In *Effective English*, edited by W. George Crouch. American Institute of Banking, New York, 1959.

"As you move up the success ladder, what you write and what you say will determine in part your rate of climb. It is neither too early nor too late to become practiced in the art of communication."

Part II

What Style Is Appropriate for Business and Technical Writing?

> *"Nothing in literature is so 'desperately mortal' as a stylish prose.
> . . . The prose of men who are intent upon their matter and write
> only to be understood tends . . . to gain from the passage of time. Its
> content, even when mediocre in itself, acquires historical interest; its
> most commonplace terms of expression grow pleasantly archaic;
> proverbs and idioms which were part of the common speech seem to
> the . . . reader to be proofs of individual genius; even characteristic
> faults are at worst refreshingly unlike the faults of our own prose.
> . . . There must be added in our own age a strong bias in favour of
> strictly functional beauty and a quite unusual distrust of rhetoric.
> Decoration, externally laid on and inorganic, is now hardly tolerated
> except in the human face."*
>
> *C. S. Lewis**

"Style" is the slippery term we use to refer to the way you write; it is the
manner with which you handle your *matter* or, in other words, the *how*
you do with your *what.* Just as your haircut, clothing, posture, man-
nerisms, and expressions blend together to comprise your personal style,
so the choice of your words, the structure and length of your sentences,
the kind of material you select, the attitude you express toward your
material and your audience, and the many other ways you deal with your
subject all blend together to comprise your writing style. And, as with
your clothes and behavior, your writing style can and should be changed
as befits the occasion. Faded jeans might be suitable at a picnic, for example. But for a formal dinner, you would want to wear a tuxedo. In a simi-

* From *English Literature in the Sixteenth Century*, p. 272. (By permission of Oxford University Press)

lar vein, a colloquial style might be appropriate in a memorandum asking a colleague to have lunch with you. But for a description of a complex mechanism, you might find a less personal style more suitable to your purpose.

If you were a poet, playwright, novelist, or essayist, you might write with the aim of entertaining your readers with the beauty and originality of your style. But as a writer in business and industry, you will usually have as your chief purpose the clear communication to your readers of facts and ideas. The style that will most likely suit your purpose in such writing is a nonliterary style—a plain style that is obedient to the careful transmission of fact and that does not call attention to itself.

Perhaps the chief trait of this utilitarian style is a clarity that is achieved through straightforward organization, precise and understandable word choice, economical sentences and paragraphs, and the use of visual devices to supplement the verbal communication. Another trait is an effective tone, or the portrayal of yourself as an inoffensive person whose facts are reliable and whose judgment is balanced. Still another trait of this style is a conciseness that comes not from sacrificing necessary information, but from eliminating unnecessary words, phrases, sentences, or even paragraphs. You may find the writing style characterized here to have a functional beauty that few poets would disown.

Among the essays that follow, John Fielden's "'What Do You Mean I Can't Write?'" describes what good writing for business and industry really is and offers an inventory that helps to improve your writing. The essays by Stuart Chase and Rudolf Flesch show you how to write effectively in workaday language, while Samuel T. Williamson's "How to Write Like a Social Scientist" pokes fun at the excess verbiage that obscures the meaning of so many written communications. Morris Freedman's essay points up stylistic obstacles often found in the way of clear communication, and John Walter's explains that the use of standard English in business and technical writing is not the only stylistic criterion by which the effectiveness of such writing is judged. The last article in Part II, S. I. Hayakawa's "How Words Change Our Lives," should make you more sensitive to language. Reading the essays in Part II will help you develop a writing style appropriate for the business world.

Selection Seven

"What Do You Mean I Can't Write?"

JOHN S. FIELDEN

John Fielden is a professor of business at Harvard University. The Written Performance Inventory he offers in his article serves as a touchstone for writers in business and industry who are concerned with an effective style.

What do businessmen answer when they are asked, "What's the most troublesome problem you have to live with?" Frequently they reply, "People just can't write! What do they learn in college now? When I was a boy . . . !"

There is no need to belabor this point; readers know well how true it is. HBR subscribers, for example, recently rated the "ability to communicate" as the prime requisite of a promotable executive (see Figure 7-1).[1] And, of all the aspects of communication, the written form is the most troublesome, if only because of its formal nature. It is received cold, without the communicator's tone of voice or gesture to help. It is rigid; it cannot be adjusted to the recipients' reactions as it is being delivered. It stays "on the record," and cannot be undone. Further, the reason it is in fact committed to paper is usually that its subject is considered too crucial or significant to be entrusted to casual, short-lived verbal form.

Businessmen know that the ability to write well is a highly valued asset in a top executive. Consequently, they become ever more conscious of their writing ability as they consider what qualities they need in order to rise in their company.

John S. Fielden, " 'What Do You Mean I Can't Write?' " *Harvard Business Review*, May–June 1964. Copyright © 1964 by the President and Fellows of Harvard College; all rights reserved.

[1] See also, C. Wilson Randle, "How to Identify Promotable Executives," HBR May–June 1956, p. 122.

Figure 7-1. Qualities that characterize promotable executives (Taken from Exhibit III, Garda W. Bowman, "What Helps or Harms Promotability?" [Problems in Review], *Harvard Business Review*, January-February 1964, p. 14.)

They know that in big business today ideas are not exchanged exclusively by word of mouth (as they might be in smaller businesses). And they know that even if they get oral approval for something they wish to do, there will be the inevitable "give me a memo on it" concluding remark that will send them back to their office to oversee the writing of a carefully documented report.

They know, too, that as they rise in their company, they will have to be able to supervise the writing of subordinates—for so many of the memos, reports, and letters written by subordinates will go out over their signature, or be passed on to others in the company and thus reflect on the caliber of work done under their supervision.

Even the new data-processing machines will not make business any less dependent on words. For while the new machines are fine for handling tabular or computative work, someone must write up an eventual analysis of the findings in the common parlance of the everyday executive.

TIME FOR ACTION

Complaints about the inability of managers to write are a very common and justifiable refrain. But the problem this article poses—and seeks to solve—is that it is of very little use to complain about something and stop right there. I think it is about time for managers to begin to do something about it. And the first step is *to define what "it"—what good business writing—really is.*

Suppose you are a young managerial aspirant who has recently been told: "You simply can't write!" What would this mean to you? Naturally, you would be hurt, disappointed, perhaps even alarmed to have your *own* nagging doubts about your writing ability put uncomfortably on the line. "Of course," you say, "I know I'm no stylist. I don't even pretend to be a literarily inclined person. But how can I improve my writing on the job? Where do I begin? Exactly what *is* wrong with my writing?" But nobody tells you in specific, meaningful terms.

Does this mean that you can't spell or punctuate or that your grammar is disastrous? Does it mean that you can't think or organize your thoughts? Or does it mean that even though you are scrupulously correct in grammar and tightly organized in your thinking, a report or letter from you is always completely unreadable; that reading it, in effect, is like trying to butt one's head through a brick wall? Or does it mean that you are so tactless and boorish in the human relations aspect of communication that your messages actually build resentment and resistance? Do you talk "down" too much or do you talk "over your reader's head"? Just what do you do wrong?

Merely being told that you can't write is so basically meaningless and so damaging to your morale that you may end up writing more ineffectually than ever before. What you need to know is: "What are the elements of good business writing? And in which of these elements am I proficient? In which do I fall down?" If only the boss could break his complaint down into a more meaningful set of components, you could begin to do something about them.

Now let's shift and assume that you are a high-ranking manager whose job it is to supervise a staff of assistants. What can you do about upgrading the writing efforts of your men? You think of the time lost by having to do reports and letters over and over before they go out, the feasibility reports which did not look so feasible after having been befogged by an ineffectual writer, the letters presented for your signature that would have infuriated the receiver had you let them be mailed. But where are you to start?

Here is where the interests of superior and subordinate meet. Unless both arrive at a common understanding, a shared vocabulary that enables them to communicate with one another about the writing jobs that need to be done, nobody is going to get very far. No oversimplified, gimmicky slogans (such as, "Every letter is a sales letter"; "Accentuate the positive, eliminate the negative"; or "Write as you speak") are going to serve this purpose. No partial view is either—whether that of the English teacher, the logician, or the social scientist—since good business writing is not just grammar, or clear thinking, or winning friends and influencing people. It is some of each, the proportion depending on the purpose.

TOTAL INVENTORY

To know what effective business writing is, we need a total inventory of all its aspects, so that:

• Top managers can say to their training people, "Are you sure our training efforts in written communications are not tackling just part of the problem? Are we covering all aspects of business writing?"

• A superior can say to an assistant, "Here, look; this is where you are weak. See? It is one thing when you write letters that you sign, another when you write letters that I sign. The position and power of the person we are writing to make a lot of difference in *what* we say and *how* we say it."

• The young manager can use the inventory as a guide to self-improvement (perhaps even ask his superior to go over his writing with him, using the writing inventory as a means of assuring a common critical vocabulary).

• The superior may himself get a few hints about how he might improve his own performance.

Such an inventory appears in Figure 7-2. Notice that it contains four basic categories—*readability*, *correctness*, *appropriateness*, and *thought*. Considerable effort has gone into making these categories (and the subtopics under them) as mutually exclusive as possible, although some overlap is inevitable. But even if they are not completely inclusive, they are still far less general than an angry, critical remark, such as, "You cannot write."

Furthermore, you should understand that these four categories are not listed in order of importance, since their importance varies according to the abilities and the duties of each individual. The same thing is true of the subtopics; I shall make no attempt to treat each of them equally, but will simply try to do some practical, commonsense highlighting. I will begin with readability, and discuss it most fully, because this is an area where half-truths abound and need to be scotched before introducing the other topics.

READABILITY

What is *readability*? Nothing more than a clear style of writing. It does not result absolutely (as some readability experts would have you believe) from mathematical counts of syllables, of sentence length, or of abstract words. These inflexible approaches to readability assume that all writing is being addressed to a general audience. Consequently, their greatest use is in forming judgments about the readability of such things as mass magazine editorial copy, newspaper communications, and elementary textbooks.

To prove this point, all you need do is to pick up a beautifully edited magazine like the *New England Journal of Medicine* and try to read an article in it. You as a layman will probably have trouble. On the other hand, your physician will tell you that the article is a masterpiece of readable exposition. But, on second look, you will still find it completely unreadable. The reason, obviously, is that you do not have the background or the vocabulary necessary to understand it. The same thing would hold true if you were to take an article from a management science quarterly, say, one dealing with return on investment or statistical decision making, and give it to the physician. Now he is likely to judge this one to be completely incomprehensible, while you may find it the most valuable and clear discussion of the topic you have ever seen.

In situations like this, it does not make much difference whether the sentences are long or short; if the reader does not have the background to understand the material, he just doesn't. And writing such specialized articles according to the mathematical readability formulas is not going to make them clearer.

Nevertheless, it is true that unnecessarily long, rambling sentences are wearing to read. Hence you will find these stylistic shortcomings mentioned

1. READABILITY

READER'S LEVEL

- [ ] Too specialized in approach
- [ ] Assumes too great a knowledge of subject
- [ ] So underestimates the reader that it belabors the obvious

SENTENCE CONSTRUCTION

- [ ] Unnecessarily long in difficult material
- [ ] Subject-verb-object word order too rarely used
- [ ] Choppy, overly simple style (in simple material)

PARAGRAPH CONSTRUCTION

- [ ] Lack of topic sentences
- [ ] Too many ideas in single paragraph
- [ ] Too long

FAMILIARITY OF WORDS

- [ ] Inappropriate jargon
- [ ] Pretentious language
- [ ] Unnecessarily abstract

READER DIRECTION

- [ ] Lack of "framing" (i.e., failure to tell the reader about purpose and direction of forthcoming discussion)
- [ ] Inadequate transitions between paragraphs
- [ ] Absence of subconclusions to summarize reader's progress at end of divisions in the discussion

FOCUS

- [ ] Unclear as to subject of communication
- [ ] Unclear as to purpose of message

2. CORRECTNESS

MECHANICS

- [ ] Shaky grammar
- [ ] Faulty punctuation

FORMAT

- [ ] Careless appearance of documents
- [ ] Failure to use accepted company form

COHERENCE

- [ ] Sentences seem awkward owing to illogical and ungrammatical yoking of unrelated ideas
- [ ] Failure to develop a logical progression of ideas through coherent, logically juxtaposed paragraphs

3. APPROPRIATENESS

A. UPWARD COMMUNICATIONS

TACT

- [ ] Failure to recognize differences in position between writer and receiver
- [ ] Impolitic tone — too brusk, argumentative, or insulting

SUPPORTING DETAIL

- [ ] Inadequate support for statements
- [ ] Too much undigested detail for busy superior

OPINION

- [ ] Adequate research but too great an intrusion of opinions
- [ ] Too few facts (and too little research) to entitle drawing of conclusions

Figure 7-2. Written performance inventory

☐ Presence of unasked for but clearly implied recommendations

ATTITUDE

☐ Too obvious a desire to please superior

☐ Too defensive in face of authority

☐ Too fearful of superior to be able to do best work

B. DOWNWARD COMMUNICATIONS

DIPLOMACY

☐ Overbearing attitude toward subordinates

☐ Insulting and/or personal references

☐ Unmindfulness that messages are representative of management group or even of company

CLARIFICATION OF DESIRES

☐ Confused, vague instructions

☐ Superior is not sure of what is wanted

☐ Withholding of information necessary to job at hand

MOTIVATIONAL ASPECTS

☐ Orders of superior seem arbitrary

☐ Superior's communications are manipulative and seemingly insincere

4. THOUGHT

PREPARATION

☐ Inadequate thought given to purpose of communication prior to its final completion

☐ Inadequate preparation or use of data known to be available

COMPETENCE

☐ Subject beyond intellectual capabilities of writer

☐ Subject beyond experience of writer

FIDELITY TO ASSIGNMENT

☐ Failure to stick to job assigned

☐ Too much made of routine assignment

☐ Too little made of assignment

ANALYSIS

☐ Superficial examination of data leading to unconscious overlooking of important pieces of evidence

☐ Failure to draw obvious conclusions from data presented

☐ Presentation of conclusions unjustified by evidence

☐ Failure to qualify tenuous assertions

☐ Failure to identify and justify assumptions used

☐ Bias, conscious or unconscious, which leads to distorted interpretation of data

PERSUASIVENESS

☐ Seems more convincing than facts warrant

☐ Seems less convincing than facts warrant

☐ Too obvious an attempt to sell ideas

☐ Lacks action-orientation and managerial viewpoint

☐ Too blunt an approach where subtlety and finesse called for

in Figure 7-2. [The trick a writer has to learn is to judge the complexity and the abstractness of the material he is dealing with, and to cut his sentences down in those areas where the going is especially difficult. It also helps to stick to a direct subject-verb-object construction in sentences wherever it is important to communicate precisely. Flights of unusually dashing style should be reserved for those sections which are quite general in nature and concrete in subject matter.]

[What about paragraphs? The importance of "paragraph construction" is often overlooked in business communication, but few things are more certain to make the heart sink than the sight of page after page of unbroken type. One old grammar book rule would be especially wise to hark back to, and that is the topic sentence. Not only does placing a topic sentence at the beginning of each paragraph make it easier for the reader to grasp the content of the communication quickly; it also serves to discipline the writer into including only one main idea in each paragraph. Naturally, when a discussion of one idea means the expenditure of hundreds (or thousands) of words, paragraphs should be divided according to subdivisions of the main idea. In fact, an almost arbitrary division of paragraphs into units of four or five sentences is usually welcomed by the reader.]

Context [As for jargon, the only people who complain about it seriously are those who do not understand it. Moreover, it is fashionable for experts in a particular field to complain about their colleagues' use of jargon, but then to turn right around and use it themselves. The reason is that jargon is no more than shop talk. And when the person being addressed fully understands this private language, it is much more economical to use it than to go through laborious explanations of every idea that could be communicated in the shorthand of jargon. Naturally, when a writer knows that his message is going to be read by persons who are not familiar with the private language of his trade, he should be sure to translate as much of the jargon as he can into common terms.]

[The same thing holds true for simplicity of language. Simplicity is, I would think, always a "good." True, there is something lost from our language when interesting but unfamiliar words are no longer used. But isn't it true that the shrines in which these antiquities should be preserved lie in the domain of poetry or the novel, and not in business communications— which, after all, are not baroque cathedrals but functional edifices by which a job can be done?]

The simplest way to say it, then, is invariably the best in business writing. But this fact the young executive does not always understand. Often he is eager to parade his vocabulary before his superiors, for fear his boss (who has never let him know that he admires simplicity, and may indeed adopt a pretentious and ponderous style himself) may think less of him.

LEADING THE READER

But perhaps the most important aspect of readability is the one listed under the subtopic "reader direction." The failure of writers to seize their reader by the nose and lead him carefully through the intricacies of his communication is like an epidemic. The job that the writer must do is to develop the "skeleton" of the document that he is preparing. And, at the very beginning of his communication, he should identify the skeletal structure of his paper; he should, in effect, frame the discussion which is to follow.

You will see many of these frames at the beginning of articles published in HBR, where the editors take great pains to tell the reader quickly what the article is about and what specific areas will come under discussion during its progress. In every business document this initial frame, this statement of purpose and direction, should appear. Furthermore, in lengthy reports there should be many such frames; indeed, most major sections of business reports should begin with a new frame.

There should also be clear transitions between paragraphs. The goal should be that of having each element in a written message bear a close relationship to those elements which have preceded and those which follow it. Frequently a section should end with a brief summary, plus a sentence or two telling the reader the new direction of the article. These rather mechanical signposts, while frequently the bane of literary stylists, are always of valuable assistance to readers.

The final aspect of readability is the category that I call "focus." This term refers to the fact that many communications seem diffuse and out of focus, much like a picture on a television screen when the antennas are not properly directed. Sometimes in a report it seems as if one report has been superimposed on another, and that there are no clear and particular points the writer is trying to make. Thus the burden is put on the reader to ferret out the truly important points from the chaos.

If a writer wants to improve the readability of his writing, he must make sure that he has thought things through sufficiently, so that he can focus his readers' attention on the salient points.

CORRECTNESS

The one thing that flies to a writer's mind when he is told he cannot write is *correctness*. He immediately starts looking for grammar and punctuation mistakes in things that he has written.

But mistakes like these are hardly the most important aspects of business writing. The majority of executives are reasonably well educated and can, with a minimum of effort, make themselves adequately proficient in the "mechanics" of writing. Furthermore, as a man rises in his company, his typing (at least) will be done by a secretary, who can (and should) take the blame if a report is poorly punctuated and incorrect in grammar, not to mention being presented in an improper "format."

Then what is the most important point? Frequently, the insecure writer allows small mistakes in grammar and punctuation to become greatly magnified, and regards them as reflections on his education and, indeed, his social acceptability. A careless use of "he don't" may seem to be as large a disgrace in his mind as if he attended the company banquet in his shorts. And in some cases this is true. But he should also realize (as Figure 7-2 shows) that the ability to write *correctly* is not synonymous with the ability to write *well*. Hence, everyone should make sure that he does not become satisfied with the rather trivial act of mastering punctuation and grammar.

It is true, of course, that, in some instances, the inability to write correctly will cause a lack of clarity. We can all think of examples where a misplaced comma has caused serious confusion—although such instances, except in contracts and other legal documents, are fortunately rather rare.

A far more important aspect of correctness is "coherence." Coherence means the proper positioning of elements within a piece of writing so that it can be read clearly and sensibly. Take one example:

Incoherent: "I think it will rain. However, no clouds are showing yet. Therefore, I will take my umbrella."

Coherent: "Although no clouds are showing, I think it will rain. Therefore, I will take my umbrella."

Once a person has mastered the art of placing related words and sentences as close as possible to each other, he will be amazed at how smooth his formerly awkward writing becomes. But that is just the beginning. He will still have to make sure that he has placed paragraphs which are related in thought next to one another, so that the ideas presented do not have to leapfrog over any intervening digressions.

APPROPRIATENESS

I have divided the category *appropriateness* into two sections reflecting the two main types of internal business communications—those going upward in the organization and those going downward. This distinction is one that cannot be found in textbooks on writing, although the ideas included here are commonplace in the human relations area.

There is an obvious difference between the type of communication that a boss writes to his subordinate and the type that the subordinate can get away with when he writes to his boss (or even the type that he drafts for his boss's signature). I suspect that many managers who have had their writing criticized had this unpleasant experience simply because of their failure to recognize the fact that messages are affected by the relative positions of the writer and the recipient in the organizational hierarchy.

UPWARD COMMUNICATIONS

Let us roughly follow the order of the subtopics included under upward communications in Figure 7-2. "Tact" is important. If a subordinate fails to recognize his role and writes in an argumentative or insulting tone, he is almost certain to reap trouble for himself (or for his boss if the document goes up under the boss's actual or implied signature). One of the perennially difficult problems facing any subordinate is how to tell a superior he is wrong. If the subordinate were the boss, most likely he *could* call a spade a spade; but since he is not, he has problems. And, in today's business world, bosses themselves spend much time figuring out how to handle problem communications with discretion. Often tender topics are best handled orally rather than in writing.

Two other subtopics—"supporting detail" and "opinion"—also require a distinction according to the writer's role. Since the communication is going upward, the writer will probably find it advisable to support his statements with considerable detail. On the other hand, he may run afoul of superiors who will be impatient if he gives too much detail and not enough generalization. Here is a classic instance where a word from above as to the amount of detail required in a particular assignment would be of inestimable value to the subordinate.

The same holds true for "opinion." In some cases, the subordinate may be criticized for introducing too many of his personal opinions—in fact, often for giving any recommendation at all. If the superior wishes the subordinate to make recommendations and to offer his own opinions, the burden is on the superior to tell him. If the superior fails to do so, the writer can at least try to make it clear where facts cease and opinions begin; then the superior can draw his own conclusions.

The writer's "attitude" is another important factor in upward communications. When a subordinate writes to his boss, it is almost impossible for him to communicate with the blandness that he might use if he were writing a letter to a friend. There may be many little things that he is doing throughout his writing that indicate either too great a desire to impress the boss or an insecurity which imparts a feeling of fearfulness, defensiveness, or truculence in the face of authority.

DOWNWARD COMMUNICATIONS

While the subordinate who writes upward in the organization must use "tact," the boss who writes down to his subordinates must use "diplomacy." If he is overbearing or insulting (even without meaning to be), he will find his effectiveness as a manager severely limited. Furthermore, it is the foolish manager who forgets that, when he communicates downward, he speaks as a representative of management or even of the entire company. Careless messages have often played an important part in strikes and other corporate human relations problems.

It is also important for the superior to make sure that he has clarified in his own mind just what it is he wishes to accomplish. If he does not, he may give confused or vague instructions. (In this event, it is unfair for him to blame a subordinate for presenting a poorly focused document in return.) Another requirement is that the superior must make sure that he has supplied any information which the subordinate needs but could not be expected to know, and that he has sufficiently explained any points which may be misleading.

Motivation is important, too. When a superior gives orders, he will find that over the long run he will not be able to rely on mere power to force compliance with his requests. It seems typically American for a subordinate to resent and resist what he considers to be arbitrary decisions made for unknown reasons. If at all possible, the superior not only should explain the reasons why he gives an order but should point out (if he can) why his decision can be interpreted as being in the best interests of those whom it affects.

I am not, however, suggesting farfetched explanations of future benefits. In the long run, those can have a boomerang effect. Straight talk, carefully and tactfully couched, is the only sensible policy. If, for example, a subordinate's request for a new assignment has been denied because he needs further experience in his present assignment, he should be told the facts. Then, if it is also true that getting more experience may prepare him for a better position in the future, there is no reason why this information should not be included to "buffer" the impact of the refusal of a new assignment.

THOUGHT

Here—a most important area—the superior has a tremendous vested interest in the reporting done by his subordinates. There is no substitute for the thought content of a communication. What good is accomplished if a message is excellent in all the other respects we have discussed—if it is readable, correct, and appropriate—yet the content is faulty? It can even do harm if the other aspects succeed in disguising the fact that it is superficial, stupid, or biased. The superior receiving it may send it up through the organization with his signature, or, equally serious, he may make an important (and disastrous) decision based on it.

Here is the real *guts* of business writing—intelligent content, something most purveyors of business writing gimmicks conveniently forget. It is also something that most training programs shortchange. The discipline of translating thoughts into words and organizing these thoughts logically has no equal as intellectual training. For there is one slogan that is true: "Disorganized, illogical writing reflects a disorganized, illogical (and untrained) mind."

That is why the first topic in this section is "preparation." Much disorganized writing results from insufficient preparation, from a failure to think through and isolate the purpose and the aim of the writing job. Most writers tend to think as they write; in fact, most of us do not even know what it is we think until we have actually written it down. The inescapability of making a well-thought-out outline before dictating seems obvious.

A primary aspect of *thought*, consequently, is the intellectual "competence" of the writer. If a report is bad merely because the subject is far beyond the experience of the writer, it is not his fault. Thus his superior should be able to reject the analysis and at the same time accept the blame for having given his assistant a job that he simply could not do. But what about the many cases where the limiting factor *is* basically the intellectual capacity of the writer? It is foolish to tell a man that he cannot *write* if in effect he simply does not have the intellectural ability to do the job that has been assigned to him.

Another aspect of thought is "fidelity to the assignment." Obviously the finest performance in the world on a topic other than the one assigned is fruitless, but such violent distortions of the assignment fortunately are rare. Not so rare, unfortunately, are reports which subtly miss the point, or wander away from it. Any consistent tendency on the part of the writer to drag in his pet remedies or favorite villains should be pointed out quickly, as should persistent efforts to grind personal axes.

Another lapse of "fidelity" is far more forgivable. This occurs when an eager subordinate tends to make too much of a routine assignment and consistently turns memos into 50-page reports. On the other hand, some subordinates may consistently make too little of an assignment and tend to do superficial and poorly researched pieces of work.

Perhaps the most important aspect of thought is the component "analysis." Here is where the highly intelligent are separated from those less gifted, and those who will dig from those who content themselves with superficial work. Often subordinates who have not had the benefit of experience under a strict taskmaster (either in school or on the job) are at a loss to understand why their reports are considered less than highly effective. Such writers, for example, may fail to draw obvious conclusions from the data that they have presented. On the other hand, they may offer conclusions which are seemingly unjustified by the evidence contained in their reports.

Another difficulty is that many young managers (and old ones, too) are unsophisticated in their appreciation of just what constitutes evidence. For example, if they base an entire report on the fact that sales are going to go up the next year simply because one assistant sales manager thinks so, they should expect to have their conclusions thrown out of court. They may also find themselves in difficulty if they fail to identify and justify assumptions which have been forced on them by the absence of factual data. Assumptions, of course, are absolutely necessary in this world of imperfect knowledge—especially when we deal with future developments—but it is the writer's responsibility to point out that certain assumptions have been made and that the validity of his analysis depends on whether or not these assumptions prove to be justified.

Another serious error in "analysis" is that of bias. Few superiors will respect a communication which is consciously or unconsciously biased. A writer who is incapable of making an objective analysis of all sides of a question, or of all alternatives to action, will certainly find his path to the top to be a dead end. On the other hand, especially in many younger writers, bias enters unconsciously, and it is only by a patient identification of the bias that the superior will be able to help the subordinate develop a truly objective analytical ability.

PERSUASIVENESS

This discussion of bias in reporting raises the question of "persuasiveness." "Every letter is a sales letter of some sort," goes the refrain. And it is true that persuasiveness in writing can range from the "con man" type of presentation to that which results from a happy blending of the four elements of business writing I have described. While it would be naive to suggest that it is not often necessary for executives to write things in manipulative ways to achieve their ends *in the short run*, it would be foolish to imply that this type of writing will be very effective with the same people (if they are reasonably intelligent) *over the long run*. Understandably, therefore, the "con man" approach will not be particularly effective in the large business organization.

On the other hand, persuasiveness is a necessary aspect of organizational writing. Yet it is difficult to describe the qualities which serve to make a communication persuasive. It could be a certain ring of conviction about the way recommendations are advanced; it could be enthusiasm, or an understanding of the reader's desires, and a playing up to them. One can persuade by hitting with the blunt edge of the axe or by cutting finely with the sharp edge to prepare the way. Persuasion could result from a fine sense of discretion, of hinting but not stating overtly things which are impolitic to mention; or it could result from an action-orientation that conveys top management's desire for results rather than a more philosophical approach to a subject. In fact, it could be many things.

In an organization, the best test to apply for the propriety of persuasiveness is to ask yourself whether you would care to take action on the basis of what your own communication presents. In the long run, it is dangerous to assume that everyone else is stupid and malleable; so, if you would be offended or damaged in the event that you were persuaded to take the action suggested, you should restate the communication. This test eliminates needless worry about slightly dishonest but well-meaning letters of congratulation, or routine progress reports written merely for a filing record, and the like. But it does bring into sharp focus those messages that cross the line from persuasiveness to bias; these are the ones that will injure others and so eventually injure you.

CONCLUSION

No one can honestly estimate the billions of dollars that are spent in U.S. industry on written communications, but the amount must be staggering. By contrast, the amount of thinking and effort that goes into improving the effectiveness of business writing is tiny—a mouse invading a continent. A written performance inventory (like Figure 7-2) in itself is not the answer. But a checklist of writing elements should enable executives to speak about writing in a common tongue and hence be a vehicle by which individual and group improvement in writing can take place.

By executives' own vote, no aspect of a manager's performance is of greater importance to his success than communication, particularly written communication. By the facts, however, no part of business practice receives less formal and intelligent attention. What this article asserts is that when an individual asks, "What do you mean I can't write?"—and has every desire to improve—his company owes him a sensible and concrete answer.

DISCUSSION AND ACTIVITY

1. Fielden stresses that unified paragraphs and simple language are crucial for a readable style in business and technical writing. When is a paragraph unified? Does simple language indicate that a writer has a limited vocabulary?

2. Fielden observes that the ability to write *correctly* is not the same as the ability to write *well*. Explain the distinction.

Selection Eight

Gobbledygook

STUART CHASE

Stuart Chase's "Gobbledygook" is now a classic in business and technical writing circles. It deflates the notion that windy, pretentious language is a component of effective style.

Said Franklin Roosevelt, in one of his early presidential speeches: "I see one-third of a nation ill-housed, ill-clad, ill-nourished." Translated into standard bureaucratic prose his statement would read:

It is evident that a substantial number of persons within the Continental boundaries of the United States have inadequate financial resources with which to purchase the products of agricultural communities and industrial establishments. It would appear that for a considerable segment of the population, possibly as much as 33.3333* of the total, there are inadequate housing facilities, and an equally significant proportion is deprived of the proper types of clothing and nutriment.

* Not carried beyond four places.

This rousing satire on gobbledygook—or talk among the bureaucrats—is adapted from a report[1] prepared by the Federal Security Agency in an attempt to break out of the verbal squirrel cage. "Gobbledygook" was coined by an exasperated Congressman, Maury Maverick of Texas, and means using two, or three, or ten words in the place of one, or using a five-syllable word where a single syllable would suffice. Maverick was censuring the forbidding

From *Power of Words*, copyright, 1953, 1954, by Stuart Chase. Reprinted by permission of Harcourt Brace Jovanovich, Inc.

[1] This and succeeding quotations from F.S.A. report by special permission of the author, Milton Hall.

prose of executive departments in Washington, but the term has now spread to windy and pretentious language in general.

"Gobbledygook" itself is a good example of the way a language grows. There was no word for the event before Maverick's invention; one had to say: "You know, that terrible, involved, polysyllabic language those government people use down in Washington." Now one word takes the place of a dozen.

A British member of Parliament, A. P. Herbert, also exasperated with bureaucratic jargon, translated Nelson's immortal phrase, "England expects every man to do his duty":

England anticipates that, as regards the current emergency, personnel will face up to the issues, and exercise appropriately the functions allocated to their respective occupational groups.

A New Zealand official made the following report after surveying a plot of ground for an athletic field.[2]

It is obvious from the difference in elevation with relation to the short depth of the property that the contour is such as to preclude any reasonable developmental potential for active recreation.

Seems the plot was too steep.

An office manager sent this memo to his chief:

Verbal contact with Mr. Blank regarding the attached notification of promotion has elicited the attached representation intimating that he prefers to decline the assignment.

Seems Mr. Blank didn't want the job.

A doctor testified at an English trial that one of the parties was suffering from "circumorbital haematoma."

Seems the party had a black eye.

In August 1952 the U.S. Department of Agriculture put out a pamphlet entitled: "Cultural and Pathogenic Variability in Single-Condial and Hyphaltip Isolates of Hemlin-Thosporium Turcicum Pass."

Seems it was about corn leaf disease.

On reaching the top of the Finsteraarhorn in 1845, M. Dollfus-Ausset, when he got his breath, exclaimed:

The soul communes in the infinite with those icy peaks which seem to have their roots in the bowels of eternity.

Seems he enjoyed the view.

[2] This item and the next two are from the piece on gobbledygook by W. E. Farbstein, New York *Times*, March 29, 1953.

A government department announced:

Voucherable expenditures necessary to provide adequate dental treatment required as adjunct to medical treatment being rendered a pay patient in in-patient status may be incurred as required at the expense of the Public Health Service.

Seems you can charge your dentist bill to the Public Health Service. Or can you?

LEGAL TALK

Gobbledygook not only flourishes in government bureaus but grows wild and lush in the law, the universities, and sometimes among the literati. Mr. Micawber was a master of gobbledygook, which he hoped would improve his fortunes. It is almost always found in offices too big for face-to-face talk. Gobbledygook can be defined as squandering words, packing a message with excess baggage and so introducing semantic "noise." Or it can be scrambling words in a message so that meaning does not come through. The directions on cans, bottles, and packages for putting the contents to use are often a good illustration. Gobbledygook must not be confused with double talk, however, for the intentions of the sender are usually honest.

I offer you a round fruit and say, "Have an orange." Not so an expert in legal phraseology, as parodied by editors of *Labor:*

I hereby give and convey to you, all and singular, my estate and interests, right, title, claim and advantages of and in said orange, together with all rind, juice, pulp and pits, and all rights and advantages therein . . . anything hereinbefore or hereinafter or in any other deed or deeds, instrument or instruments of whatever nature or kind whatsoever, to the contrary, in any wise, notwithstanding.

The state of Ohio, after five years of work, has redrafted its legal code in modern English, eliminating 4,500 sections and doubtless a blizzard of "whereases" and "hereinafters." Legal terms of necessity must be closely tied to their referents but the early solons tried to do this the hard way, by adding synonyms. They hoped to trap the physical event in a net of words, but instead they created a mumbo-jumbo beyond the power of the layman, and even many a lawyer, to translate. Legal talk is studded with tautologies, such as "cease and desist," "give and convey," "irrelevant, incompetent, and immaterial." Furthermore, legal jargon is a dead language; it is not spoken and it is not growing. An official of one of the big insurance companies calls their branch of it "bafflegab." Here is a sample from his collection.[3]

[3] Interview with Clifford B. Reeves by Sylvia F. Porter, New York *Evening Post*, March 14, 1952.

One-half to his mother, if living, if not to his father, and one-half to his
mother-in-law, if living, if not to his mother, if living, if not to his father.
Thereafter payment is to be made in a single sum to his brothers. On the
one-half payable to his mother, if living, if not to his father, he does not
bring in his mother-in-law as the next payee to receive, although on the
one-half to his mother-in-law, he does bring in the mother or father.

You apply for an insurance policy, pass the tests, and instead of a
straightforward "here is your policy," you receive something like this:

This policy is issued in consideration of the application therefor, copy of
which application is attached hereto and made part hereof, and of the
payment for said insurance on the life of the above-named insured.

ACADEMIC TALK

The pedagogues may be less repetitious than the lawyers, but many use even
longer words. It is a symbol of their calling to prefer Greek and Latin
derivatives to Anglo-Saxon. Thus instead of saying: "I like short clear
words," many a professor would think it more seemly to say: "I prefer an
abbreviated phraseology, distinguished for its lucidity." Your professor is
sometimes right, the longer word may carry the meaning better—but not
because it is long. Allen Upward in his book *The New Word* warmly advocates
Anglo-Saxon English as against what he calls "Mediterranean" English, with
its polysyllables built up like a skyscraper.

Professional pedagogy, still alternating between the Middle Ages and
modern science, can produce what Henshaw Ward once called the most
repellent prose known to man. It takes an iron will to read as much as a page
of it. Here is a sample of what is known in some quarters as "pedagese":

Realization has grown that the curriculum or the experiences of learners
change and improve only as those who are most directly involved examine their goals, improve their understandings and increase their skill in
performing the tasks necessary to reach newly defined goals. This places
the focus upon teacher, lay citizen and learner as partners in curricular
improvement and as the individuals who must change, if there is to be
curriculum change.

I think there is an idea concealed here somewhere. I think it means: "If we
are going to change the curriculum, teacher, parent, and student must all
help." The reader is invited to get out his semantic decoder and check on my
translation. Observe there is no technical language in this gem of pedagese,
beyond possibly the word "curriculum." It is just a simple idea heavily
oververbalized.

In another kind of academic talk the author may display his learning to
conceal a lack of ideas. A bright instructor, for instance, in need of prestige

may select a common sense proposition for the subject of a learned monograph—say, "Modern cities are hard to live in" and adorn it with imposing polysyllables: "Urban existence in the perpendicular declivities of megalopolis . . ." et cetera. He coins some new terms to transfix the reader—"megadecibel" or "strato-cosmopolis"—and works them vigorously. He is careful to add a page or two of differential equations to show the "scatter." And then he publishes, with 147 footnotes and a bibliography to knock your eye out. If the authorities are dozing, it can be worth an associate professorship.

While we are on the campus, however, we must not forget that the technical language of the natural sciences and some terms in the social sciences, forbidding as they may sound to the layman, are quite necessary. Without them, specialists could not communicate what they find. Trouble arises when experts expect the uninitiated to understand the words; when they tell the jury, for instance, that the defendant is suffering from "circumorbital haematoma."

Here are two authentic quotations. Which was written by a distinguished modern author, and which by a patient in a mental hospital? You will find the answer at the end of the article.

Have just been to supper. Did not knowing what the woodchuck sent me here. How when the blue blue blue on the said anyone can do it that tries. Such is the presidential candidate.

No history of a family to close with those and close. Never shall he be alone to be alone to be alone to be alone to be alone to lend a hand and leave it left and wasted.

REDUCING THE GOBBLE

As government and business offices grow larger, the need for doing something about gobbledygook increases. Fortunately the biggest office in the world is working hard to reduce it. The Federal Security Agency in Washington,[4] with nearly 100 million clients on its books, began analyzing its communication lines some years ago, with gratifying results. Surveys find trouble in three main areas: correspondence with clients about their social security problems, office memos, official reports.

Clarity and brevity, as well as common humanity, are urgently needed in this vast establishment which deals with disability, old age, and unemploy-

[4] Now the Department of Health, Education, and Welfare.

ment. The surveys found instead many cases of long-windedness, foggy meanings, clichés, and singsong phrases, and gross neglect of the reader's point of view. Rather than talking to a real person, the writer was talking to himself. "We often write like a man walking on stilts."

Here is a typical case of long-windedness:

Gobbledygook as found: "We are wondering if sufficient time has passed so that you are in a position to indicate whether favorable action may now be taken on our recommendation for the reclassification of Mrs. Blank, junior clerk-stenographer, CAF 2, to assistant clerk-stenographer, CAF 3?"

Suggested improvements: "Have you yet been able to act on our recommendation to reclassify Mrs. Blank?

Another case:

Although the Central Efficiency Rating Committee recognizes that there are many desirable changes that could be made in the present efficiency rating system in order to make it more realistic and more workable than it now is, this committee is of the opinion that no further change should be made in the present system during the current year. Because of conditions prevailing throughout the country and the resultant turnover in personnel, and difficulty in administering the Federal programs, further mechanical improvement in the present rating system would require staff retraining and other administrative expense which would seem best withheld until the official termination of hostilities, and until restoration of regular operations.

The F.S.A. invites us to squeeze the gobbledygook out of this statement. Here is my attempt:

The Central Efficiency Rating Committee recognizes that desirable changes could be made in the present system. We believe, however, that no change should be attempted until the war is over.

This cuts the statement from 111 to 30 words, about one-quarter of the original, but perhaps the reader can do still better. What of importance have I left out?

Sometimes in a book which I am reading for information—not for literary pleasure—I run a pencil through the surplus words. Often I can cut a section to half its length with an improvement in clarity. Magazines like *The Reader's Digest* have reduced this process to an art. Are long-windedness and obscurity a cultural lag from the days when writing was reserved for priests and cloistered scholars? The more words and the deeper the mystery, the greater their prestige and the firmer the hold on their jobs. And the better the candidate's chance today to have his doctoral thesis accepted.

The F.S.A. surveys found that a great deal of writing was obscure although not necessarily prolix. Here is a letter sent to more than 100,000 inquirers, a classic example of murky prose. To clarify it, one needs to *add* words, not cut them:

In order to be fully insured, an individual must have earned $50 or more
in covered employment for as many quarters of coverage as half the
calendar quarters elapsing between 1936 and the quarter in which he
reaches age 65 or dies, whichever first occurs.

Probably no one without the technical jargon of the office could translate this: nevertheless, it was sent out to drive clients mad for seven years. One poor fellow wrote back: "I am no longer in covered employment. I have an outside job now."

Many words and phrases in officialese seem to come out automatically, as if from lower centers of the brain. In this standardized prose people never *get jobs*, they "secure employment"; *before* and *after* become "prior to" and "subsequent to"; one does not *do*, one "performs"; nobody *knows* a thing, he is "fully cognizant"; one never *says*, he "indicates." A great favorite at present is "implement."

Some charming boners occur in this talking-in-one's-sleep. For instance:

The problem of extending coverage to all employees, regardless of size, is
not as simple as surface appearances indicate.

Though the proportions of all males and females in ages 16–45 are essentially the same . . .

Dairy cattle, usually and commonly embraced in dairying . . .

In its manual to employees, the F.S.A. suggests the following:

Instead of	Use
give consideration to	consider
make inquiry regarding	inquire
is of the opinion	believes
comes into conflict with	conflicts
information which is of a confidential nature	confidential information

Professional or office gobbledygook often arises from using the passive rather than the active voice. Instead of looking you in the eye, as it were, and writing "This act requires . . ." the office worker looks out of the window and writes: "It is required by this statute that. . . ." When the bureau chief says, "We expect Congress to cut your budget," the message is only too clear; but usually he says, "It is expected that the departmental budget estimates will be reduced by Congress."

Gobbled: "All letters prepared for the signature of the Administrator will be single spaced."

Ungobbled: "Single space all letters for the Administrator." (Thus cutting 13 words to 7.)

ONLY PEOPLE CAN READ

The F.S.A. surveys pick up the point . . . that human communication involves a listener as well as a speaker. Only people can read, though a lot of writing seems to be addressed to beings in outer space. To whom are you talking? The sender of the officialese message often forgets the chap on the other end of the line.

A woman with two small children wrote the F.S.A. asking what she should do about payments, as her husband had lost his memory. "If he never gets able to work," she said, "and stays in an institution would I be able to draw any benefits? . . . I don't know how I am going to live and raise my children since he is disable to work. Please give me some information. . . ."

To this human appeal, she received a shattering blast of gobbledygook, beginning, "State unemployment compensation laws do not provide any benefits for sick or disabled individuals . . . in order to qualify an individual must have a certain number of quarters of coverage . . ." et cetera, et cetera. Certainly if the writer had been thinking about the poor woman he would not have dragged in unessential material about old-age insurance. If he had pictured a mother without means to care for her children, he would have told her where she might get help—from the local office which handles aid to dependent children, for instance.

Gobbledygook of this kind would largely evaporate if we thought of our messages as two way—in the above case, if we pictured ourselves talking on the doorstep of a shabby house to a woman with two children tugging at her skirts, who in her distress does not know which way to turn.

RESULTS OF THE SURVEY

The F.S.A. survey showed that office documents could be cut 20 to 50 percent, with an improvement in clarity and a great saving to taxpayers in paper and payrolls.

A handbook was prepared and distributed to key officials. They read it, thought about it, and presently began calling section meetings to discuss gobbledygook. More booklets were ordered, and the local output of documents began to improve. A Correspondence Review Section was established as a kind of laboratory to test murky messages. A supervisor could send up samples for analysis and suggestions. The handbook is now used for training new members; and many employees keep it on their desks along with the dictionary. Outside the Bureau some 25,000 copies have been sold (at 20 cents each) to individuals, governments, business firms, all over the world. It is now used officially in the Veterans Administration and in the Department of Agriculture.

The handbook makes clear the enormous amount of gobbledygook which automatically spreads in any large office, together with ways and means to keep it under control. I would guess that at least half of all the words circulating around the bureaus of the world are "irrelevant, incompetent, and immaterial"—to use a favorite legalism; or are just plain "unnecessary"—to ungobble it.

My favorite story of removing the gobble from gobbledygook concerns the Bureau of Standards at Washington. I have told it before but perhaps the reader will forgive the repetition. A New York plumber wrote the Bureau that he had found hydrochloric acid fine for cleaning drains, and was it harmless? Washington replied: "The efficacy of hydrochloric acid is indisputable, but the chlorine residue is incompatible with metallic permanence."

The plumber wrote back that he was mighty glad the Bureau agreed with him. The Bureau replied with a note of alarm: "We cannot assume responsibility for the production of toxic and noxious residues with hydrochloric acid, and suggest that you use an alternate procedure." The plumber was happy to learn that the Bureau still agreed with him.

Whereupon Washington exploded: "Don't use hydrochloric acid; it eats hell out of the pipes!"

Note: The second quotation on page 66 comes from Gertrude Stein's *Lucy Church Amiably*.

DISCUSSION AND ACTIVITY

1. What impression do you form of writers whose communications are larded with gobbledygook? Is your image the one the writers want you to form?

2. In Part I of this book, several articles pointed to reader accommodation as the primary trait of good business and technical writing. Explain how Chase's article helps you develop a writing style that accommodates your readers.

Selection Nine

Write the Way You Talk

RUDOLF FLESCH

Rudolf Flesch is a well-known author of works that teach people how to write, speak, and think effectively. In the following article, he recommends writing in a conversational style.

Ninety-nine percent of the people who come to my writing classes were born nonwriters and have stayed that way all their lives. For them, writing has always been an unpleasant chore; answering a simple letter looms ahead like a visit to the dentist. But they have to do a certain amount of writing in their careers. And knowing their writing was poor, they decided to do something about it.

No doubt when you think about improving your writing you think of grammar, rhetoric, composition—all those dull things you learned year after year in school. But most likely, these things are not your problem. You probably have a pretty good grip on these essentials. What you need is instruction in the basic principles of professional writing.

Why professional writing? Because you now write as you did in school, unconsciously trying to please the teacher by following the rules of "English composition." You're not really writing a letter to the addressee, or a report for your vice president. The pros—magazine writers, newspapermen, novelists, people who write for a living—learned long ago that they must use "spoken" English and avoid "written" English like the plague.

"Write the Way You Talk" as it appeared in *Reader's Digest*, August 1973. Condensed from *Say What You Mean* by Rudolf Flesch. Copyright © 1972 by Rudolf Flesch. By permission of Harper & Row, Publishers.

TALK ON PAPER   The secret to more effective writing is simple: *talk* to your reader. Pretend the person who'll read your letter or report is sitting across from you, or that you are on the phone with him. Be informal. Relax. Talk in your ordinary voice, your ordinary manner, vocabulary, accent and expression. You wouldn't say "Please be advised," or "We wish to inform you." Instead, something like, "You see, it's like this," or "Let me explain this." One helpful trick is to imagine yourself talking to your reader across a table at lunch. Punctuate your sentences, in your mind, with a bite from a sandwich. Intersperse your thoughts with an occasional "you know," or the person's name.

So talk—talk on paper. Go over what you've written. Does it look and sound like talk? If not, change it until it does.

USE CONTRACTIONS FREELY   There's nothing more important for improving your writing style. Use of *don't* and *it's* and *haven't* and *there's* is the No. 1 style device of modern professional writing. Once you've learned this basic trick, you can start producing prose that will be clear, informal and effective.

Take the standard opening phrase: "Enclosed please find." What's a better way of saying that? Simply, "Here's"!*

LEAVE OUT THE WORD "THAT" WHENEVER POSSIBLE   You can often omit it without changing the meaning at all. Take this sentence: "We suggest that you send us your passbook once a year." Now strike out *that*. Isn't this better and smoother? Again, this is something we do all the time in speaking.

And while you're crossing out *thats*, also go on a *which* hunt. For some reason people think *which* is a more elegant pronoun. Wrong. Usually you . can replace *which* by *that*, or leave it out altogether—and you'll get a better, more fluent, more "spoken" sentence.

USE DIRECT QUESTIONS   A conversation is not one-sided. One person speaks, then the other interrupts, often with a question, like "Really?" or "Then what?" A conversation without questions is almost inconceivable. So use a question whenever there's an opportunity, and your writing will sound more like talk.

You don't have to go out of your way to do this. Look at what you write and you'll find indirect questions—beginning with *whether*—all over the place. "Please determine whether payment against these receipts will be in order." No good. Make it: "Can we pay against these receipts? Please find out and let us know."

Or take another sentence: "Your questions and comments are invited." Again, this is really a question: "Do you have any questions or comments? If

---

* Though most of my examples are taken from business correspondence, the principles apply to *all* types of writing.

so, please let us know." There's nothing like a direct question to get some feedback.

USE PERSONAL PRONOUNS   A speaker uses *I, we* and *you* incessantly—they're part of the give-and-take of conversation. Everybody, it seems, who writes for a company or organization clings desperately to the passive voice and avoids taking the slightest responsibility. He doesn't say *we*, never says *I*, and he even avoids using the straightforward *you.* So we find phrases like "It is assumed . . . ," "it will be seen . . . ," "it is recommended. . . ." Or sentences like: "An investigation is being made and upon its completion a report will be furnished you." Instead, write, "We've made an investigation and we'll furnish you a report."

Normally, when writing for an organization, there isn't too much opportunity to say "I." But do use "I" whenever you express feelings and thoughts that are your own. Often it's better to say "I'm sorry," or "I'm pleased," than "we're sorry" or "we're pleased." And call the addressee *you.* The idea is to make your writing as personal as possible.

IT'S ALL RIGHT TO PUT PREPOSITIONS AT THE END   For 50 years, English-language experts have unanimously insisted that a preposition at the end is fine and dandy. H. W. Fowler, in *A Dictionary of Modern English Usage*, 1926, defends it enthusiastically and cites examples from Shakespeare and the Bible to Thackeray and Kipling. Yet schoolteachers still tell pupils they should never commit such a wicked crime.

Put the preposition at the end whenever it sounds right to do so. Instead of "The claimant is not entitled to the benefits for which he applied," write "The claimant isn't entitled to the benefits he applied for." Remember, grammatical superstitions are something to get rid of.

SPILL THE BEANS   There's a natural tendency in all of us to begin at the beginning and go on to the end. When you write a letter, it's the easiest way to organize your material. The trouble is, it's hard on the reader. He has a problem, or a question, and wants to know whether the answer is yes or no. If he has to wait until you're willing to tell him, his impatience and subconscious resentment will increase with every word. Rather than stumbling your way through some awkward introduction, start right in with the most important thing you want to get across. Plunge right in.

Re: . . .

Gentlemen:

In reference to the above collection item, which you instructed us to hold at the disposal of the beneficiary, we wish to advise that Mr. Ling has not called on us, nor have we received any inquiries on his behalf.

The above information is provided to you in the event you wish to give us any further instructions in the matter.

Cross out everything up to the words "Mr. Ling." Then the letter becomes (with a few other minor changes):

Re: . . .

Gentlemen:

Mr. Ling hasn't called on us, nor have we had any inquiries on his behalf. Do you have any further instructions?

You see what this does? Once the unnecessary verbiage is cleared away, the letter becomes downright elegant.

WRITE SHORT, SNAPPY SENTENCES   The ordinary reader can take in only so many words before his eyes come to a brief rest at a period. If a sentence has more than 40 words, chances are he's been unable to take in the full meaning. So break those long sentences apart, 20 words at most. It's usually quite easy to see where one idea leaves off and another begins. Then try writing *really* short sentences every so often, and watch your letters and reports wake from their customary torpor.

USE SHORT WORDS   Long, pompous words are a curse, a curtain that comes between writer and reader. Here are some familiar sayings as they would appear in a business letter. "In the event that initially you fail to succeed, endeavor, endeavor again." "All is well that terminates well."

Everybody has his own pet pomposities. Banish them from your vocabulary. Replace *locate* with *find: prior to* with *before; sufficient* with *enough; in the event that* with *if*. After those simple substitutions, weed out such other words as *determine, facilitate* and *require* whenever they show up. You'll find that it's possible to live without them. And you'll learn to appreciate the joys of simple language.

WRITE FOR PEOPLE   By far the most important thing is to give your letters just the right human touch. Express your natural feelings. If it's good news, say you're glad; if it's bad news, say you're sorry. Be as courteous, polite and interested as you'd be if the addressee sat in front of you. Some human being will read your letter and, consciously or unconsciously, be annoyed if it is cold, pleased if you're courteous and friendly.

A bank got a letter from a customer who'd moved from New York to Bermuda. He wrote to make new arrangements about his account. The bank's answer started: "We thank you for your letter advising us of your change of address." Now really! How stony and unfeeling can you get? I would at least have said something like "I noted your new address with envy."

You'll find there are rewards for improving your written work. This is the age of large organizations where it's easier to catch the eye of a superior by what you write than by what you say or do. Write the way I suggest and your stuff will stand out. Beyond the material rewards are more personal ones. When you write a particularly crisp, elegant paragraph, or a letter that conveys your thoughts clearly and simply, you'll feel a flow of creative achievement. Treasure it. It's something you've earned.

DISCUSSION AND ACTIVITY

1. If, as many people argue, technical writing is "objective," is a conversational style necessarily inappropriate for technical writing?

2. Is a conversational style appropriate for all writing in business and industry? Enumerate instances when it is not.

Selection Ten

How to Write Like a Social Scientist

SAMUEL T. WILLIAMSON

The late Samuel T. Williamson was an editor of Newsweek. *In his article, you will find six facetious rules leading to a writing style he calls "pedantic Choctaw."*

During my years as an editor, I have seen probably hundreds of job applicants who were either just out of college or in their senior years. All wanted "to write." Many brought letters from their teachers. But I do not recall one letter announcing that its bearer could write what he wished to say with clarity and directness, with economy of words, and with pleasing variety of sentence structure.

Most of these young men and women could not write plain English. Apparently their noses had not been rubbed in the drudgery of putting one simple, well-chosen word behind the other. If this was true of teachers' pets, what about the rest? What about those going into business and industry? Or those going into professions? What about those who remain at college—first for a Master of Arts degree, then an instructorship combined with work for a Ph.D., then perhaps an assistant professorship, next a full professorship and finally, as an academic crown of laurel, appointment as head of a department or as dean of a faculty?

Certainly, faculty members of a front-rank university should be better able to express themselves than those they teach. Assume that those in the English department have this ability: Can the same be said of the social scientists—economists, sociologists, and authorities on government? We need today as we never needed so urgently before all the understanding they can give us of

From *Saturday Review*, October 4, 1947. Reprinted by permission of the publisher.

problems of earning a living, caring for our fellows, and governing ourselves. Too many of them, I find, can't write as well as their students.

I am still convalescing from over-exposure some time ago to products of the academic mind. One of the foundations engaged me to edit the manuscripts of a socio-economic research report designed for the thoughtful citizen as well as for the specialist. My expectations were not high—no deathless prose, merely a sturdy, no-nonsense report of explorers into the wilderness of statistics and half-known fact. I knew from experience that economic necessity compels many a professional writer to be a cream-skimmer and a gatherer of easily obtainable material; for unless his publisher will stand the extra cost, he cannot afford the exhaustive investigation which endowed research makes possible. Although I did not expect fine writing from a trained, professional researcher, I did assume that a careful fact-finder would write carefully.

And so, anticipating no literary treat, I plunged into the forest of words of my first manuscript. My weapons were a sturdy eraser and several batteries of sharpened pencils. My armor was a thesaurus. And if I should become lost, a near-by public library was a landmark, and the Encyclopedia of Social Sciences on its reference shelves was an ever-ready guide.

Instead of big trees, I found underbrush. Cutting through involved, lumbering sentences was bad enough, but the real chore was removal of the burdocks of excess verbiage which clung to the manuscript. Nothing was big or large; in my author's lexicon, it was "substantial." When he meant "much," he wrote "to a substantially high degree." If some event took place in the early 1920's, he put it "in the early part of the decade of the twenties." And instead of "that depends," my author wrote, "any answer to this question must bear in mind certain peculiar characteristics of the industry."

So it went for 30,000 words. The pile of verbal burdocks grew— sometimes twelve words from a twenty-word sentence. The shortened version of 20,000 words was perhaps no more thrilling than the original report; but it was terser and crisper. It took less time to read and it could be understood quicker. That was all I could do. As S. S. McClure once said to me, "An editor can improve a manuscript, but he cannot put in what isn't there."

I did not know the author I was editing; after what I did to his copy, it may be just as well that we have not met. Aside from his cat-chasing-its-own-tail verbosity, he was a competent enough workman. Apparently he is well thought of. He has his doctorate, he is a trained researcher and a pupil of an eminent professor. He has held a number of fellowships and he has performed competently several jobs of economic research. But, after this long academic preparation for what was to be a life work, it is a mystery why so little attention was given to acquiring use of simple English.

Later, when I encountered other manuscripts, I found I had been too hard on this promising Ph.D. Tone-deaf as he was to words, his report was a lighthouse of clarity among the chapters turned in by his so-called academic betters. These brethren—and sister'n—who contributed the remainder of the foundation's study were professors and assistant professors in our foremost colleges and universities. The names of one or two are occasionally in newspaper headlines. All of them had, as the professorial term has it, "published."

Anyone who edits copy, regardless of whether it is good or bad, discovers in a manuscript certain pet phrases, little quirks of style and other individual traits of its author. But in the series I edited, all twenty reports read alike. Their words would be found in any English dictionary, grammar was beyond criticism, but long passages in these reports demanded not editing but actual translation. For hours at a time, I floundered in brier patches like this: "In eliminating wage changes due to purely transitory conditions, collective bargaining has eliminated one of the important causes of industrial conflict, for changes under such conditions are almost always followed by a reaction when normal conditions appear."

I am not picking on my little group of social scientists. They are merely members of a caste; they are so used to taking in each other's literary washing that it has become a habit for them to clothe their thoughts in the same smothering verbal garments. Nor are they any worse than most of their colleagues, for example:

In the long run, developments in transportation, housing, optimum size of plant, etc., might tend to induce an industrial and demographic pattern similar to the one that consciousness of vulnerability would dictate. Such a tendency might be advanced more effectively if the causes of urbanization had been carefully studied.

Such pedantic Choctaw may be all right as a sort of code language or shorthand of social science to circulate among initiates, but its perpetrators have no right to impose it on others. The tragedy is that its users appear to be under the impression that it is good English usage.

Father, forgive them; for they know not what they do! There once was a time when everyday folk spoke one language, and learned men wrote another. It was called the Dark Ages. The world is in such a state that we may return to the Dark Ages if we do not acquire wisdom. If social scientists have answers to our problems yet feel under no obligation to make themselves understood, then we laymen must learn their language. This may take some practice, but practice should become perfect by following six simple rules of the guild of social science writers. Examples which I give are sound and well tested; they come from manuscripts I edited.

RULE 1   *Never use a short word when you can think of a long one.* Never say "now" but "currently." It is not "soon" but "presently." You did not have "enough" but a "sufficiency." Never do you come to the "end" but to the "termination." This rule is basic.

RULE 2   *Never use one word when you can use two or more.* Eschew "probably." Write, "it is improbable," and raise this to "it is not improbable." Then you'll be able to parlay "probably" into "available evidence would tend to indicate that it is not unreasonable to suppose."

RULE 3   *Put one-syllable thought into polysyllabic terms.* Instead of observing that a work force might be bigger and better, write, "In addition to quantitative enlargement, it is not improbable that there is need also for qualitative improvement in the personnel of the service." If you have discovered that musicians out of practice can't hold jobs, report that "the fact of rapid deterioration of musical skill when not in use soon converts the employed into the unemployable." Resist the impulse to say that much men's clothing is machine made. Put it thus: "Nearly all operations in the industry lend themselves to performance by machine, and all grades of men's clothing sold in significant quantity involve a very substantial amount of machine work."

RULE 4   *Put the obvious in terms of the unintelligible.* When you write that "the product of the activity of janitors is expended in the identical locality in which that activity takes place," your lay reader is in for a time of it. After an hour's puzzlement, he may conclude that janitors' sweepings are thrown on the town dump. See what you can do with this: "Each article sent to the cleaner is handled separately." You become a member of the guild in good standing if you put it like this: "Within the cleaning plant proper the business of the industry involves several well-defined processes, which, from the economic point of view, may be characterized simply by saying that most of them require separate handling of each individual garment or piece of material to be cleaned."

RULE 5   *Announce what you are going to say before you say it.* This pitcher's wind-up technique before hurling towards—not at—home plate has two varieties. First is the quick wind-up: "In the following sections the policies of the administration will be considered." Then you become strong enough for the contortionist wind-up: "Perhaps more important, therefore, than the question of what standards are in a particular case, there are the questions of the extent of observance of these standards and the methods of their enforcement." Also, you can play with reversing Rule 5 and *say what you have said after you have said it.*

RULE 6   *Defend your style as "scientific."* Look down on—not up to—clear, simple English. Sneer at it as "popular." Scorn it as "journalistic." Explain your failure to put more mental sweat into your writing on the ground that "the social scientists who want to be scientific believe that we can have scientific description of human behavior and trustworthy predictions in the scientific sense only as we build adequate taxonomic systems for observable phenomena and symbolic systems for the manipulation of ideal and abstract entities."

For this explanation I am indebted to Lyman Bryson in an *SRL* [*Saturday Review of Literature*] article (Oct. 13, 1945) "Writers: Enemies of Social Science." Standing on ground considerably of his own choosing, Mr. Bryson argued against judging social science writing by literary standards.

Social scientists are not criticized because they are not literary artists. The trouble with social science does not lie in its special vocabulary. Those words are doubtless chosen with great care. The trouble is that too few social scientists take enough care with words outside their special vocabularies.

It is not too much to expect that teachers should be more competent in the art of explanation than those they teach. Teachers of social sciences diligently try to acquire knowledge; too few of them exert themselves enough to impart it intelligently.

Too long has this been excused as "the academic mind." It should be called by what it is: intellectual laziness and grubby-mindedness.

DISCUSSION AND ACTIVITY

1. Although the manuscripts Williamson edited used words "found in any English dictionary" and "grammar that was beyond criticism," long passages nevertheless required translation. Explain why writing that is "correct" might still require editing.

2. Translate the example of "pedantic Choctaw" that begins, "In the long run, developments. . . ."

Selection Eleven

The Seven Sins of Technical Writing

MORRIS FREEDMAN

*Morris Freedman is a writer, editor, and English professor at the
University of Maryland. Here he discusses seven sins of technical
writing and how to avoid committing them.*

Let me start by saying at once that I do not come to you . . . just as a
professor of English, for, frankly, I do not think that I would have very much
to say to you only as someone expert in the history of the use—and
misuse—of the language. And any remarks on literature might be confusing,
at least without extensive elaboration, for the values and objectives of literature seem so very different at first from those of technical writing—although
fundamentally many of these values and objectives coincide. And I am sure
that you are more than familiar with such things as clichés, comma splices,
fragmentary sentences, and the other abominations we deal with in freshman
composition. These obviously have nothing to do specifically with technical
writing.

But I want to say, before anyone thinks that I class technical writing
entirely by itself, immune from rules and requirements of communication
that govern other kinds of writing, that technical writing calls for the same
kind of attention and must be judged by the same standards as any other kind
of writing; *indeed, it calls for a greater attention and for higher standards*. And I say
this as a former science and medical writer for the popular press; as a former
writer of procedure manuals and directives for the government; as a former
editor of technical studies in sociology, statistics, law, and psychology; as a

former general magazine editor; as a writer of fiction, essays, and scholarly articles; and, not least, as a professor of English. We can see at once why technical writing must be measured by higher standards, or, at least, by different ones, if anyone will not grant me that they are higher. Technical writing is so immediately functional. Confusing directions accompanying an essential device in a jet plane may result in disaster; bad writing elsewhere can have as its most extreme effect merely boredom.

Yet, while technical writing implicitly calls for great care, it differs from other kinds of writing in that its practitioners are, by and large, first technicians and only incidentally writers. And principally because of this arrangement, I think, technical writing has become characterized by a collection of sins peculiar to this discipline alone. I say the *collection* is peculiar to technical writing, not any one of the sins alone. Any newspaper, weekly magazine, encyclopedia, textbook, any piece of writing you might name, will contain one or another of these sins, in greater or lesser profusion. But I know of no kind of writing that contains as many different sins in such great number as technical writing, and with such great potential for danger. To repeat, the sins in the world at large—at least, of the sort I'm talking about—often don't matter much. And sometimes, too, they don't matter in technical writing. As my students argue when I correct them in informative writing: "You got the meaning, didn't you?" Yes, I did, and so do we all get the meaning when a newspaper, a magazine, a set of directions stammers out its message. And I suppose, too, we could travel by ox-cart, or dress in burlap, or drive around with rattling fenders, and still get through a day.

But technical writing in this age can no more afford widespread sloppiness of expression, confusion of meaning, rattle-trap construction than a supersonic missile can afford to be made of the wrong materials, or be put together haphazardly with screws jutting out here and there, or have wiring circuits that may go off any way at all, or—have a self-destructive system that fails because of some fault along the way in construction. Technical writing today—as I need hardly reiterate to this audience—if it is much less than perfect in its streamlining and design may well result in machines that are less than trim, and in operation that is not exactly neat. This is at worst; at best, poor technical writing, when its effect is minimized by careful reading, hinders efficiency, wastes time. Let me remark too that the commission of any one of these sins, and of any one of many, many lesser ones, is really not likely alone to be fatal, just as one loose screw by itself is not likely to destroy a machine; but always, we know, sins come in bunches, the sin of avarice often links hands with the sin of gluttony, one loose screw may mean others, and, anyway, the ideal of no sins at all—especially in something like technical writing, where the pain of self-denial should be minimal—is always to be strived for.

A final word before I launch into the sins (whose parade, so long delayed, will prove, I hope, so much more edifying—like a medieval tableau). The seven I list might be described as cardinal ones, and as such they are broad and overlapping, perhaps, rather than specific and very clearly distinguished from one another. They all contribute to making technical writing less clear, concise, coherent, and correct than it should be.

Sin 1, then, might be described as that of *Indifference,* neglecting the reader. I do not mean anything so simple as writing down to an engineer or physicist, although this is all too common and may be considered part of this sin. This writing down—elaborating the obvious—is one reason the abstract or summary has become so indispensable a part of technical reports; very often, it is all the expert needs to read of the whole report, the rest being a matter of all too obvious detailing. Nor do I mean writing above the heads of your audience either, which is a defect likely to be taken care of by a thoughtful editor. Both writing over or under the heads of your reader, or to the side, are really matters of careless aiming and, as such, of indifference, too. But what I mean here by indifference are shortcuts of expression, elliptical diction, sloppy organization, bringing up points and letting them hang unresolved, improper or inadequate labelling of graphic material, and the like. This is communication by gutturals, grunts, shrugs, as though it were not worth the trouble to articulate carefully, as though the reader didn't matter—or didn't exist. This is basically an attitude of disrespect: *Caveat lector*—let the reader beware. Let the reader do his own work; the writer isn't going to help him.

Here is the concluding sentence from a quite respectable report, one most carefully edited and indeed presented as a model in a handbook for technical writers used by a great chemical firm. The sentence is relatively good, for it takes only a second reading to work out its meaning (perhaps only a slow first one for someone trained in reading this kind of writing):

When it is assumed that all of the cellulose is converted to ethyl cellulose, reaction conversion of cellulose to ethyl cellulose, per cent of cellulose reacted, and reaction yield of ethyl cellulose based on cellulose are each equal to 100%.

This is admittedly a tough sentence to get across simply, considering that "cellulose" is repeated in several different contexts. Yet two guiding principles would have made it much clearer: (1) always put for your reader first things first (here, the meaning hangs on the final phrase, "each equal to 100%," which comes at the end of a complicated series); and (2) clearly separate items in a series. (The second rule seems to me one of the most important in technical writing where so many things have to be listed so often.) Here is the recast sentence:

If all the cellulose is converted to ethyl cellulose, each of the following factors is then equal to 100%:
  1. reaction conversion of cellulose to ethyl cellulose.

2. proportion of cellulose reacted.
3. reaction yield of ethyl cellulose based on cellulose.

The changes are not great, certainly, but in the process we have eliminated the indisputable notion of a percent being equal to a percent, and have arranged the series so that both the eye and the mind together can grasp the information immediately. Sin 1 then can be handled, one way, by cutting out indirect Rube Goldbergish contraptions and hitting your points directly on their heads, one, two, three.

The remaining sins I shall discuss are extensions of this primal one, disregard for the reader. Sin 2 may be designated as *Fuzziness*, that is, a general fuzziness of communication—vague words, meaningless words, wrong ones. The reader uses his own experience to supply the meaning in such writing; the writing itself acts only as a collection of clues. The military specializes in this sort of thing. I recall an eerie warning in an air force mess hall: "Anyone smoking in or around this mess hall will be dealt with accordingly." It still haunts me. Here is a caution in a handbook of technical writing with which you may be familiar: "Flowery, euphemistic protestations of gratitude are inappropriate." We know what this means, of course, but we ourselves supply the exact meaning. It happens that a "euphemism" is "the substitution of an inoffensive or mild expression for one that may offend or suggest something unpleasant." At least, that's what *Webster's Collegiate* says it is.

Here are some other examples: "The intrinsic labyrinth of wires must be first disentangled." The writer meant "network," not "labyrinth"; and I think he meant "internal" for "intrinsic" and "untangled" for "disentangled." Item: "The liquid contents of the container should then be disgorged via the spout by the operator." Translation: "The operator should then empty the container." Here is a final long one:

When the element numbered one is brought into tactual contact with the element numbered two, when the appropriate conditions of temperature have been met above the previously determined safety point, then there will be exhibited a tendency for the appropriate circuit to be closed and consequently to serve the purpose of activating an audible warning device.

Translation:

When the heat rises above the set safety point, element one touches element two, closing a circuit and setting off a bell.

Prescription to avoid Sin 2: use concrete, specific words and phrases whenever you can, and use only those words whose meaning you are sure of. (A dictionary, by the way, is only a partial help in determining the correct and *idiomatic* use of a word.) English is perhaps the richest of languages in offering a variety of alternatives for saying the same thing.

Sin 3 might be called the sin of *Emptiness*. It is the use of jargon and big words, pretentious ones, where perfectly appropriate and acceptable small and normal words are available. (There is nothing wrong with big words in themselves, provided they are the best ones for the job. A steam shovel is right for moving a boulder, ridiculous for picking up a handkerchief.) We may want to connect this sin with the larger, more universal one of pride, the general desire to seem important and impressive. During World War II a high government official devoted much time to composing an effective warning for a sticker to be put above light switches. He emerged with "Illumination is required to be extinguished on these premises on the termination of daily activities," or something of the sort. He meant "Put the lights out when you go home."

The jargon I'm talking about is not the technical language you use normally and necessarily for efficient communication. I have in mind only the use of a big word or a jumble of words for something that can be said more efficiently with familiar words and straightforward expressions. I have in mind also a kind of code language used to show that you're an insider, somewhere or other: "Production-wise, that's a high-type machine that can be used to finalize procedure. The organization is enthused." There is rarely any functional justification for saying "utilize" or "utilization" for "use," "prior to" for "before," "the answer is in the affirmative or negative" for "yes or no," or for using any of the "operators, or false verbal limbs," as George Orwell called them, like "render inoperative," "prove unacceptable," "exhibit a tendency to," "serve the purpose of," and so on and on.

Again, one can handle this sin simply by overcoming a reluctance to saying things directly; the most complex things in the world can be said in simple words, often of one syllable. Consider propositions in higher math or logic, the Supreme Court decisions of men like Brandeis and Holmes, the poetry of Shakespeare. I cannot resist quoting here Sir Arthur Quiller-Couch's rendition in jargon of Hamlet's "To be or not to be, that is the question." I am sure you all know the full jargon rendition of the soliloquy. "To be, or the contrary? Whether the former or the latter be preferable would seem to admit of some difference of opinion."

Sin 4 is an extension of 3: just plain *Wordiness*. The principle here is that if you can say anything with more words than necessary for the job, then by all means do so. I've already cited examples of this sin above, but compounded with other sins. Here is a purer example, the opening of a sentence in a technical writing handbook: "Material to be contained on the cover of the technical report includes. . . ." This can be reduced to "The cover of the technical report should include. . . ." Another example, less pure: "The front-mounted blade of the bulldozer is employed for earth moving operations on road construction jobs." Translation: "The bulldozer's front blade moves earth in road building." Item: "There is another way of accomplishing this purpose, and that is by evaporation." Translation: "Evaporation is another way of doing this." Instead of saying simply that "the bulldozer's front blade moves earth," you say it "is employed for earth moving operations," throwing in "employed" and "operations," as though "moves" alone is too weak to do this tremendous job. The cure for this sin? Simply reverse the mechanism: say what you have to in the fewest words.

Sin 5, once again an extension of the immediately preceding sin, is a matter of *Bad Habits*, the use of pat phrases, awkward expressions, confusing sentence structure, that have, unfortunately, become second nature. Again, I'm not alluding to the perfectly natural use of familiar technical expressions, which may literally be called clichés, but which are not efficiently replaceable. Sin 5 is a matter of just not paying attention to what you say, with the result that when you do suddenly pay attention, you see the pointlessness or even humor of what you have set down. Perhaps the most common example of this sin is what has been called "deadwood," or what may be called "writing for the simple minded." Examples: "red in color," "three in number," "square in shape," "the month of January," "the year 1956," "ten miles in distance," and the like. What else is red but a color, three but a number, square but a shape, January but a month, 1956 but a year, ten miles but a distance? To say that something is "two inches wide and three inches long" is to assume that your reader can't figure out length and width from the simple dimensions "two inches by three inches." I once read that a certain machine was 18 feet high, "vertically," the writer made sure to add; and another time that a certain knob should be turned "right, in direction."

A caution is needed here. There are many obvious instances when qualification is necessary. To say that something is "light," for example, is plainly mysterious unless you add "in color" or "in weight" or, perhaps, "in density" (unless the context makes such addition "deadwood").

I would include under Sin 5 the locutions "as far as that is concerned" (lately shortened to "as far as that"), "as regards," "with regard to," "in the case of" ("In the case of the case enclosing the instrument, the case is being studied"). These are all too often just lazy ways of making transitions (and, thus, incidentally, quite justifiable when speed of writing is a factor).

Sin 6 is the *Deadly Passive*, or, better, deadening passive; it takes the life out of writing, making everything impersonal, eternal, remote and dead. The deadly passive is guaranteed to make any reading matter more difficult to understand, to get through, and to retain. Textbook writers in certain fields have long ago learned to use the deadly passive to create difficulties where none exist; this makes their subject seem weightier, and their accomplishment more impressive. (And, of course, if this is ever what you have in mind on an assignment, then by all means use the deadly passive.) Sin 6 is rarely found alone; it is almost indispensable for fully carrying out the sins of wordiness and jargon. Frequently, of course, the passive is not a sin and not deadly, for there simply is no active agent and the material must be put impersonally.

Examples of this sin are so easy to come by, it is difficult to find one better than another. Here is a relatively mild example of Sin 6.

The standardization of procedure in print finishing can be a very important factor in the efficient production of service pictures. In so far as possible, the smallest number of types and sizes of paper should be employed, and the recommended processing followed. The fewer paper grades and processing procedures used, the fewer errors and make-overs that are likely. Make-overs are time-consuming and costly.

Here it is with the deadly passive out and some other changes made:

To produce service pictures efficiently, a standard way of finishing prints can be very important. You should use as few types and sizes of paper as possible, and you should follow the recommended procedure for processing. In this way, you will make fewer errors, and have to re-do less work. You save time and money.

Associated with the deadly passive, as you might see from the two passages above, is the use of abstract nouns and adjectives for verbs. Verbs always live; nouns and adjectives just sit there, and abstract nouns aren't even there. Of course, there are a number of other ways of undoing the passivity of the passage I quoted, and of making other improvements, just as there were other ways of handling any of the specimens I have cited in the train of horrors accompanying my pageant of sins.

Finally we come to Sin 7, the one considered the deadliest by many, and not only by teachers of English but by technical writers and technologists of various sorts: *Mechanical Errors.* I don't think this sin the deadliest of all. It does happen to be the easiest one to recognize, the one easiest to deal with "quantitatively," so to speak, and the easiest one to resist. I suppose it is considered deadliest because then those who avoid it can so quickly feel virtuous. It can promptly be handled by good words alone. Actually most technical writing happens to be mechanically impeccable; not one of the examples I have used tonight had very much mechanically wrong with it. If anything, technical people tend to make too much of formal mechanisms. I remember working with a physicist who had much trouble saying anything in writing. While his general incapacity to write was almost total, one thing he did know, and know firmly, and that was that a split infinitive was to be abhorred. That, and using a preposition to end a sentence with. He could never communicate the simplest notion coherently, but he never split an infinitive or left a preposition at the end of a sentence. If Nobel Prizes were to be awarded for never splitting infinitives or for encapsulating prepositions within sentences, he would be a leading candidate.

There are a handful of mechanical errors which are relevant to technical writing, and these are important because they are so common, especially in combination with other sins. (Split infinitives or sentence-ending prepositions, need I say, are not among them.) These are dangling participles and other types of poorly placed modifiers, and ambiguous references. There are others, a good number of others, but the ones I mention creep in most insidiously and most often.

Here are some examples stripped down to emphasize the errors:

Raising the temperature, the thermostat failed to function.

Who or what raised the temperature? Not the thermostat, I presume; and if it did somehow, as the result of current flowing in its wiring, then this ought to be said quite plainly.

The apparatus is inappropriately situated in the corner since it is too small.

What is too small? Apparatus or corner?

Every element in the device must not be considered to be subject to ab-normal stress.

What is meant here is that "Not every element in the apparatus must be considered subject to abnormal stress," almost the opposite of the original taken literally.

I should like to conclude by emphasizing something I glanced at in my introduction, that the seven sins of technical writing are to be avoided not so much by a specific awareness of each, accompanied by specific penance for each, as by a much more general awareness, by an attitude toward subject matter, writing process, and reader that can best be described only as "respectful." You will not help yourself very much if you rely on such purely mechanical aids as Rudolf Flesch's formulas[1] for "readable writing," or on slide rules measuring readability, much as you may be tempted to do so. These can be devil's snares, ways to make you think you are avoiding sin. There are no general texts, either, at present that will help you in more than very minor ways. The only aids you can safely depend on are the good book itself, that is, a good dictionary (there are many poor ones), any of the several volumes by H. W. Fowler, and occasional essays, here and there, by George Orwell, Jacques Barzun, Herbert Read, Somerset Maugham, and others. And these, I stress, can only be *aids*. What is most important in eliminating sin in technical writing is general attitude—as it may well be in eliminating sin anywhere.

I repeat that technical writing must be as rationally shaped as a technical object. A piece of technical writing, after all, is something that is shaped into being for a special purpose, much as a technical object. The design engineer should be guided in his work by the requirements of function almost alone. (Of course, if he happens to have a boss who likes to embellish the object with

[1] In 1943, Rudolf Flesch introduced a readability formula based on three factors: average sentence length, relative number of affixed morphemes (prefixes, suffixes, inflectional endings), and relative number of personal references.

useless doo-dads, why then he may have to modify his work accordingly to keep his job—as automobile designers do every day; but we try never to have in mind unreasonable situations of this sort.) It is as pointless for the designer engineer to use three bolts where one would do (both for safety and function), to make an object square when its use dictates it should be round, to take the long way through a process when there is a short way, as it is for the technical writer to commit any of the sins I have mentioned. Technical writing— informative writing of any sort—should be as clean, as functional, as inevitable as any modern machine designed to do a job well. If I will not be misunderstood in throwing out this thought, I should like to suggest to you that good technical writing should be like good poetry—every word in its exact place for maximum effect, no word readily replaceable by another, not a word too many or too few, and the whole combination, so to speak, invisible, not calling attention to its structure, seemingly effortless, perfectly adapted to its subject.

If one takes this general approach to the shaping of a piece of technical writing, and there really can't be much excuse for any other, then there is no need to worry about any of the sins I mention. Virtue may not come at once or automatically, for good writing never comes without effort, however fine one's intention, but it will certainly come, and perhaps even bring with it that same satisfaction the creative engineer experiences. Technical writing cleansed of its sins is no less worthy, no less impressive, an enterprise than good engineering itself. Like mathematics to physics, technical writing is a handmaid to technology, but like mathematics, too, it can be a helpmate, that is, an equal partner. But it can achieve this reward of virtue only by emphasizing the virtues of writing equally with those of technology.

DISCUSSION AND ACTIVITY

1. In Sin 3, Freedman observes that the jargon he objects to "is not the technical language you use normally and necessarily for efficient communication." What he objects to is "the use of a big word or a jumble of words for something that can be said more efficiently with familiar words and straightforward expressions." What are some terms peculiar to your specialty that would make communication more efficient if you write for other specialists? Would the same terms work if you were writing for nonspecialists?

2. Can you think of a writing situation in which the use of the passive voice is permissible? Consult your English handbook to see if it cites instances when the passive voice might be more appropriate than the active.

Selection Twelve

Usage and Style in Technical Writing: A Realistic Position

JOHN A. WALTER

John Walter is president of the Association of Teachers of Technical Writing and professor of English at the University of Texas, Austin. His article gives you a historical view of usage and style in writing.

All of you here have more than an academic interest in usage and style, the two aspects of technical writing I am to discuss with you today. After all, a concern for both is a daily necessity for technical writers and editors everywhere. I also assume that you know a good deal about both topics, dealing with them as you do in the day-by-day routine of your work. Moreover, all of you have doubtless been exposed to formal training in English, and training in English, both at the secondary school and the college levels, inevitably involves one with some attention to usage and style—at some level of sophistication or another.

Sophistication—perhaps I can take that as the key word here. At any rate, mention of it leads me to confess that I have no way of knowing what assumption can be fairly made about the extent of knowledge any group of technical writers may have on this subject of usage and style, nor can I know how refined that knowledge may be. If I seem to belabor the obvious in what I shall have to say to you—or to be engaged in flogging deceased equines, I hope you will be tolerant. I can only take my cue from my own experience as a student of language and as a technical editor, as well as from my informal observations of the state of knowledge on this subject among those of my fellow members of the Society of Technical Writers and Publishers with whom I have had occasion to discuss the matter. In any case, I propose to

Reprinted with permission from John A. Walter, Fellow, Society for Technical Communication.

discuss briefly some of the language developments of the last thirty or forty years that have had a direct bearing on current thinking with respect to usage and style, outline as best I can the stand taken today on the subject by the best known scholars in this field of language study, and mention some of the best and most readily available published studies on the subject, mention some of the special problems of style, and talk a little on the implications of some of the newer linguistic work in a study of style.

A revolution in language study has been underway for the last several decades, a revolution that has seen the development of new grammars of English and attendant new attitudes toward matters of usage and style. Like all revolutions, this one has involved—is involving, I should say—the over-throw of an older regime. In this case, the older regime is what linguists call school grammar—the kind of grammar you and I were taught in the public schools. This school grammar, together with its pronouncements on usage and style, had its origin in the first published grammars in English, those published during the 18th century. Notable among them were Bishop Robert Lowth's *A Short Introduction to English Grammar* (1762), the English progenitor of most school grammars, and Lindley Murray's *English Grammar* (1802), the American seminal work. Both of these works continue to exert influence today. As classical scholars, Lowth and his imitators took as their model for English grammar the Latin grammar with which they were familiar, and they formulated English grammar to correspond with that of Latin, ignoring what present-day linguists take as an axiomatic truth, the necessity of study-ing a language as it is actually spoken and written. Moreover, Lowth and his followers took the position that it was their right and responsibility to prescribe what speakers and writers *should* say and write rather than report or describe how language is used by competent, respected users. Along with the prescription for correct usage, the early grammarians also condemned those usages of which they did not personally approve. Thus it was, for instance, that rules against the split infinitive and the terminal preposition came into being. In Latin the infinitive is a single word, of course, and consequently it would be ridiculous to "split" it. In Latin a preposition at the end of a sentence would not be acceptable. And so it went—a proliferation of rules and proscriptions, most of them having their roots in Latin but some having their origin in the personal tastes of the grammarians themselves. When some brave souls objected, even in those early days, that some of the best writers (Shakespeare, Chaucer, Milton, for example) were guilty of "errors," the grammarians were not in the least abashed and solemnly declared that the greats would have been even greater had they been more attentive to "cor-rectness."

These early grammarians were profoundly influential, at least in the schools; and as I mentioned earlier, texts used in the schools today still bear witness to their influence. Our children are still taught an unyielding, single standard of correctness, they are still led to believe that there exists some nebulous thing called "good English," but they have the uneasy feeling that they do not possess it and that they will never attain it. And if they are alert enough, they may note from time to time that not even their English teachers speak "good" English in unguarded moments.

Beginning roughly with the work of Leonard Bloomfield, whose *Language* was published in 1933, linguists have turned their backs on the tradition of school grammars and have sought to develop new studies of language based on close study of English as it is actually spoken and written. They began with objections to the Latin origin of school grammars, attacks on the semantic basis of the definitions in the older grammars that proved their inaccuracy and inadequacy (parts of speech, etc.), objected to the emphasis on written English in those works and insisted on the primacy of spoken language over written, arguing that the spoken language preceded written by centuries, and most particularly—and of most concern to us—they objected to the inflexible stand on usage and correctness promulgated by the school grammars. Perhaps I should add at this point that not all of the older grammars were entirely discredited by the new linguists; the monumental works of Jesperson, Kruisinga, and Poutsma, to mention but a few of the greats, continue to deserve and hold the respect of contemporary scholars. The work of these men was descriptive rather than prescriptive or proscriptive.

We do not have time, of course, for even a cursory account of the new grammars—the structuralist or the transformational-generative varieties, of such men as Fries, Nelson Francis, Trager and Smith for the first or of such men as Noam Chomsky, the chief theoretician of the transformationalists for the second. The point here is that these revolutionaries have brought about an entirely new attitude toward usage and correctness and, to a lesser extent, have stimulated new approaches to the study of English prose style.

The school grammarians, if not the traditionalists, tended to assume a single standard of correctness or acceptability in matters of usage, and they used Latin as their model and writing as their corpus for study. They ignored, in their books at least, the quite obvious fact that no one uses the same kind of language in speech as he uses in formal writing. They devoted much space to elaborating the requirements of a variety of English tenses in a variety of modes (ignoring that the verb forms themselves remained remarkably constant, with tense variation having to be shown by auxiliaries), gave extensive treatment to the proprieties of usage with such auxiliaries as *shall* and *will*, ignoring that speakers, at least, simply say *I'll, you'll, he'll,* etc., not knowing or caring whether the *'ll* represented *shall* or *will,* etc. Lest I be misunderstood, let me add hastily that I am well aware of the necessity of differentiating these forms in some technical writing—specifications and legal papers, for instance. The school grammarians devoted much time to case forms, and schoolmarms continue to try (in class, at least) to stamp out *It's me* and give artificial respiration to the moribund if not completely extinct *whom*. Time was spent on the logic of language—with its analogy to algebra and the silly argument that the double negative is not permissible because two negatives make a positive, although it did not seem to occur to those who advanced this argument that if a double negative makes a positive, then a triple negative must make a negative again. Other concerns had (and have) to do with subject-verb agreement when the grammatical subject is at odds with the real subject, as with some collectives, subjects of amount or quantity, and the like; the so-called group genitive; adjective comparison and adverbial forms (go *slow*, for instance); the so-called dangling construction; matters of subordination and coordination; pronoun problems; and so on.

The new linguist's position (especially the structuralist's) was simple: he said, in effect, let's find out how speakers and writers actually use their language. And, of course, they found at once that no single standard exists— that speakers and writers use language differently depending on a large number of variables: occasion or situation, purpose, age and sex of audience, geographic location, and so on. So they were led to recognition that there is no one such thing as good usage but rather that there are usages and usages. Ultimately they formulated statements which attempted to define the acceptable varieties or levels of usage. Paul Roberts put it very well when he said that what we call correctness is "relative to time, place, and circumstance and is really a sociological matter rather than a linguistic matter. Correctness is an accident of history, which bestowed power and prestige on certain groups and not on others, which made one city the capital of a country and not another. There are no linguistic reasons for preferring 'Jim and I saw it' to 'Jim and me seen it' though there are very powerful social ones."

Thus linguists saw acceptable English, to begin with, as what is called a prestige dialect—socially desirable but not necessarily more grammatical or more effective as communication than usages universally condemned.

There came into being, then, the concept of levels or varieties of usage, with two broad categories being "standard" and "non-standard." Within the category of standard are numerous subcategories, depending to start with on whether we are concerned with speech or writing. Within the realm of writing there are, as we all know, several varieties: formal (this includes much if not most technical writing), informal, and what Porter Perrin calls "general" English, acceptable for unlimited use. Moreover, these broad categories are not sharply distinguishable; they tend to shade off into one another. The criteria for choice in usage is no longer simply whether a form is correct but rather it is a blending of judgments according to a number of criteria: (1) is it appropriate to the purpose of the communication? (2) is it appropriate to the subject and the situation? (3) is it appropriate to the listener or reader?—in terms of clarity, acceptability, and liveliness? (4) is it appropriate to the speaker or writer? As Robert Pooley says, "Good English is that form of speech [or writing] which is appropriate to the purpose of the speaker, true to the language as it is, and comfortable to the speaker and listener. It is the product of custom, neither cramped by rule nor freed from all restraint; it is never fixed, but changes with the organic life of the language." In short, good usage is—and always will be—relative rather than absolute.

Such a position on usage is obviously liberal rather than conservative. It is not, as some opponents have angrily asserted, a purely permissive attitude that maintains that anything that *is* is all right, but it is certainly more tolerant than the position of those purists whose usage and grammatical roots are still in the 18th century.

Who are the chief spokesmen for this current thinking on usage? They are many, too many to name here, but those whose work you can readily gain access to are not so numerous: Robert Pooley's *English Usage* (1946), Margaret Bryant's *Current American Usage* (1961), Perrin's *Index to English* (1966), Bergen and Cornelia Evans' *Dictionary of Contemporary American Usage* (1957) are the best and most readily available. (I should emphasize that these are books on usage which deal with practical problems and are thus helpful to the writer and editor; they do not present accounts of the new grammars I referred to earlier in this discussion.) Perhaps I should add that the 1961 Third International G & C Merriam Dictionary is also useful though it does not, of course, provide the lively and interesting discussions afforded by the other works listed.

Now what about style? What, if any, effects has the new linguistics had on this important aspect of writing? Let's come to an answer to this question in a moment. First, let me say that important as style obviously is, I have found few discussions of the subject that were at all helpful to my students—or to me, for that matter. Aphoristic definitions abound: style is the man; it is proper words in proper places; the dress of thoughts; the way you write; etc. But what does the writer aim for?

Most of us were exhorted in school to aim for clarity, emphasis, force, grace, coherence, etc. But how? I'm afraid we can do nothing with such general statements about style—except, perhaps, admire the excellence of the style in which they are expressed. In this realm of general advice, I believe the statement of Jacques Barzun and Henry Graff is as good as any. They say, "you cannot aim directly at style, at clarity, precision, and all the rest; you can only remove the many possible obstacles to understanding, while preserving as much as you can of your spontaneous utterance. . . . Clarity comes when others can follow; coherence when thoughts hang together; logic when their sequence is valid." And then they go on to say, "To the general public 'revise' is a noble word and 'tinker' is a trivial one, but to the writer the difference between them is only the difference between the details of hard work and the effect it achieves. The successful revision of a . . . manuscript is made up of an appalling number of small, local alterations. Rewriting is nothing but able tinkering."

What have the new linguists contributed on this subject of style? So far, very little. But they have pointed the way and made some beginnings toward some interesting possibilities that further exploration may bring to fruition. Modern linguists, for instance, point out that style is a matter of "linguistic choice," and as we all know very well, there is always a variety—sometimes a bewildering variety—of ways a thing may be said. Modern transformationalist grammar with its rules for the generation of sentences and its rules for transforming these basic sentences into different sorts of statements helps to focus attention—and light—on the implications of the choices available. Richard Ohmann, for instance, has published some interesting speculations on the possible promise of transformational grammatical analysis in determining the nature of style and in determining ways of bringing off a certain style. H. A. Gleason in his recent *Linguistics and English Grammar* devotes considerable space to the implications of new linguistics in the study of style, and he tentatively advances some examples of the direction such work may take. His discussion of clarity, redundancy, and ambiguity, for instance, is quite thought provoking, and in particular his treatment of what he calls "front-heaviness," "excessive embedding," and "poor transitions" (stemming from transformational terminology, in the main) is as illuminating as anything I have seen on the subject of effective sentence structuring.

Another contemporary student of language, a colleague of mine named James Sledd (author of *A Short Introduction to English Grammar*—a title he borrowed from Bishop Lowth), devotes a useful portion of his book to style, and his treatment of the subject of subordination and coordination is especially interesting. He demolishes—to my satisfaction, at least—the traditional notion that the most important thought of a sentence *must* be contained in what we call the main clause. For example, consider "He said that you are thoroughly incompetent." I don't suppose any of us would be so tradition-bound by our early studies of grammar and syntax as to insist that "He said" expresses the main idea in that sentence. Francis Christensen has also presented some provocative ideas in his *Notes Toward a New Rhetoric* (1967), and recent issues of *College Composition and Communication* have carried articles by a number of writers who have had ideas stemming from their study of the new grammar.

But style is more than a matter of generating sentences in a mechanical or inspirational way—and particularly is this true of the work of technical writers and editors. Style, especially for the technical editor, is a rhetorical matter too, and we must be concerned with those recurrent offenses against effective style which so frequently plague us. In my own experience as an editor, I have found—and recorded—a great many of them, and have, indeed, attempted from time to time to classify them and see whether they exhibit some sort of pattern that would make it possible to deal with them in a more effective way. As time permits, now, I should like to list some of the problems I seem to encounter over and over again, though I shall have to limit myself to a mere mention and at the word or phrase level alone at that.

At this limited level, then, I would name the following types of difficulty: (1) confusion of terms, (2) misuse of terms, (3) overuse of certain terms and phrases, (4) redundant use of terms and phrases, (5) completely superfluous use of certain terms, (6) pointless use of vague and meaningless adverbs and adjectives, (7) too much use of jargon and shoptalk, and (8) an overfondness for impressive or big words and phrases. Examples of each of these categories and identification of some of the subspecies within categories are numerous, and I can only hope that mention of a few of each will be sufficiently indicative of their nature that you will be able to expand the list from your own experience.

I shall not have time to discuss the problem with the items listed, but reference to such books as Bergen and Cornelia Evans' *Dictionary* . . . , not to mention Webster's, will easily clarify the problem.

In the first category, then, confusion of terms, appear such pairs as the following: ability/capacity, adjust/arrival, anxious/eager, assure/insure, constriction/restriction, continual/continuous, deteriorate/degenerate, effective/effectual, evidence/proof, oral/verbal, perfect/unique, practical/practicable, principal/principle, and many, many others.

In the second category, misused terms, appear such items as about/approximately, administrate [many backformations are misused], anticipate/expect, appreciate/understand, balance/remainder, better than/more than, climatic/climactic, corrode/ erode, encounter/ experience, experimentation/ experiment, filtrate/filter, majority/most, preventative/preventive, proportion/part, reaction/opinion, respectfully/respectively, theory/idea, etc.

Third are the overused terms and phrases, of which I can only list a few: activate, approach (as a noun), around (for about or nearly), assist, basis, blueprint (for plan, policy), category (for class), cognizant, character (for sort or kind), commence, communicate, contact, cross-section, demonstrate, description (for kind or sort), develop, effort (for work), endeavor, essential, facility, facilitate, feasible, function, implement (as a verb), inform, initiate, materialize, minimize, optimum, personnel, philosophy (for plan), sufficient, terminate, transmit, utilize and utilization, etc.

Redundancy in technical writing is probably one of its worst faults, and the following are typical: absolutely essential, actual experience, adequate enough, aluminum metal, assemble together, close proximity, collaborate together, completely eliminated, consensus of opinion, couple together, final completion, follow after, hopeful optimism, important essentials, mutual cooperation, only alternative, original source, past history, protrude out, refer back, etc. A special type of redundant phrasing occurs when verbs are piled up—when one appears in noun form—as in the following: provides continuous indication of, make a calibration, make a purchase, prepare a cost estimate, perform the measurement, functions to transmit, perform the lubrication operation, operates to correct, etc.

Completely superfluous words may be found in the following phrases: gasket located inside the/hexagonal in shape/minute droplets/may result in damage to/numerals are used to identify/range all the way from/throughout the entire/uniformly consistent/two cubic feet in volume/limited in scope/cylindrical in shape/earlier in date/few in number/first initiated into/formed by virtue of/between the limits of A and B/cancel out/etc.

In the category of vague and meaningless adverbs and adjectives are these: evidently/ negligible/ suitable/ fairly/ rather/ reasonably/ relatively/ comparatively/ approximately/ various/ definitely/ nearly/ considerably/ excessive/ sufficient/undue/substantial/significant/appreciably/and the like. It is not that these are bad terms, of course; it is that they are so often used without being tied down to anything definite. We can scarcely know whether something is relatively difficult or expensive without knowing what the relation is to.

The use of jargon or shoptalk is inevitable—and perfectly appropriate—when we are talking or writing to readers who understand it—in other words, to those who share the idiom. But it is easy to fall into the habit of using such language in communicating with those to whom it is completely unfamiliar. A few examples follow: breadboard, call out (for refer), state-of-the-art, peaks (for "reach a maximum"), ruggedize, tote board, monkey (not in reference to the animal), pessimize, megs, pot (for potentiometer), etc. Every field of professional activity produces its own special slang or shoptalk; the important consideration for the writer is whether his reader will understand.

Finally, we may mention some representative examples of the fondness for using "big" words, pretentious terms, when readily available and commonly known simple terms exist which will do the job just as well. Consider the following: abatement/abrogate/ameliorate/aperature/fabricate/facilitate/formulate/generate/gravitate/illumination/cessation/cognizant/implement/predilection/ propensity/ refractory/ scrutinize/ terminate/ ruination (!)/ unequivocal/ utilization/ vindication/ etc. The list of pretentious words could, of course, be extended almost endlessly, but our objective should be, in writing or editing, to use the simplest word consistent with accuracy.

I have not intended to imply, in listing offensive words and phrases in the foregoing categories, that good technical writing style is altogether a matter of choosing suitable words or avoiding unsuitable ones. Many, many other factors are involved in good style, of course, but I shall have to content myself with this brief effort at specificity.

What we find, then, in a review of current thinking on usage and style is that, first of all, there does not exist an absolute standard. The technical writer must keep loose and flexible, as current slang might put it. But he must keep alert and observant. He should read widely and take note of usages accepted by respected writers; he should keep up with as many publications as he can on the subject of usage and style. English never stays put; it is alive, growing, changing, and often in a state of controversial flux. The editor and writer must not, therefore, allow his views and convictions to ossify, to crystallize into rock-like prejudice. This does not mean, however, that all of the so-called rules we learned in school are now passé; most of them are still pretty valid. But rules of usage are like laws in science: they merely reflect what exists—they do not determine what exists.

DISCUSSION AND ACTIVITY

1. How does descriptive grammar differ from prescriptive or proscriptive grammar? How does descriptive grammar affect our attitude toward usage and style in writing? Does descriptive grammar tolerate complete permissiveness in usage and style? If not, why not?

2. In Walter's lists of confused terms and misused terms, explain the difference between each pair of terms separated by slashes.

Selection Thirteen

How Words Change Our Lives

S. I. HAYAKAWA

> *U.S. Senator S. I. Hayakawa is a semanticist and a former president of San Francisco State University. His study of how we react to words should help writers choose words more carefully.*

The end product of education, yours and mine and everybody's, is the total pattern of reactions and possible reactions we have inside ourselves. If you did not have within you at this moment the pattern of reactions which we call "the ability to read English," you would see here only meaningless black marks on paper. Because of the trained patterns of response, you are (or are not) stirred to patriotism by martial music, your feelings of reverence are aroused by the symbols of your religion, you listen more respectfully to the health advice of someone who has "M.D." after his name than to that of someone who hasn't. What I call here a "pattern of reactions," then, is the sum total of the ways we act in response to events, to words and to symbols.

Our reaction patterns—our semantic habits, as we may call them—are the internal and most important residue of whatever years of education or miseducation we may have received from our parents' conduct toward us in childhood as well as their teachings, from the formal education we may have had, from all the sermons and lectures we have listened to, from the radio programs and the movies and television shows we have experienced, from all the books and newspapers and comic strips we have read, from the conversations we have had with friends and associates, and from all our experiences. If, as the result of all these influences that make us what we are, our semantic habits are reasonably similar to those of most people around us, we are regarded as "well-adjusted," or "normal," and perhaps "dull." If our semantic habits are noticeably different from those of others, we are regarded as "individualistic" or "original," or, if the differences are disapproved of or viewed with alarm, as "screwball" or "crazy."

Semantics is sometimes defined in dictionaries as "the science of the meaning of words"—which would not be a bad definition if people didn't assume that the search for the meanings of words begins and ends with looking them up in a dictionary.

If one stops to think for a moment, it is clear that to define a word, as a dictionary does, is simply to explain the word with more words. To be thorough about defining, we should next have to define the words used in the definition, then define the words used in defining the words used in the definition . . . and so on. Defining words with more words, in short, gets us at once into what mathematicians call an "infinite regress." Alternatively, it can get us into the kind of run-around we sometimes encounter when we look up "impertinence" and find it defined as "impudence," so we look up "impudence" and find it defined as "impertinence." Yet—and here we come to another common reaction pattern—people often act as if words can be explained fully with more words. To a person who asked for a definition of jazz, Louis Armstrong is said to have replied, "Man, when you got to ask what it is, you'll never get to know," proving himself to be an intuitive semanticist as well as a great trumpet player.

WHAT SEMANTICS IS ABOUT

Semantics, then, does not deal with the "meaning of words" as that expression is commonly understood. P. W. Bridgman, the Nobel-prize winner and physicist, once wrote, "The true meaning of a term is to be found by observing what a man does with it, not by what he says about it." He made an enormous contribution to science by showing that the meaning of a scientific term lies in the operations, the things done, that establish its validity, rather than in verbal definitions.

Here is a simple, everyday kind of example of "operational" criticism. If you say, "This table measures six feet in length," you could prove it by taking a foot rule, performing the operation of laying it end to end while counting, "One . . . two . . . three . . . four . . ." But if you say—and revolutionists have started uprisings with just this statement— "Man is born free, but everywhere he is in chains!"—what operations could you perform to demonstrate its accuracy or inaccuracy?

But let us carry this suggestion of "operationalism" outside the physical sciences where Bridgman applied it, and observe what "operations" people perform as the result of both the language they use and the language other people use in communicating to them. Here is a personnel manager studying an application blank. He comes to the words "Education: Harvard University," and drops the application blank in the wastebasket (that's the "operation") because, as he would say if you asked him, "I don't like Harvard men." This is an instance of "meaning" at work—but it is not a meaning that can be found in dictionaries.

If I seem to be taking a long time to explain what semantics is about, it is because I am trying, in the course of explanation, to introduce the reader to a certain way of looking at human behavior. Semantics—especially the general semantics of Alfred Korzybski (1879–1950), Polish-American scientist and educator—pays particular attention not to words in themselves, but to semantic reactions—that is, human responses to symbols, signs and symbol-systems, including language.

I say *human* responses because, so far as we know, human beings are the only creatures that have, over and above that biological equipment which we have in common with other creatures, the additional capacity for manufacturing symbols and systems of symbols. When we react to a flag, we are not reacting simply to a piece of cloth, but to the meaning with which it has been symbolically endowed. When we react to a word, we are not reacting to a set of sounds, but to the meaning with which that set of sounds has been symbolically endowed.

A basic idea in general semantics, therefore, is that the meaning of words (or other symbols) is not in the words, but in our own semantic reactions. If I were to tell a shockingly obscene story in Arabic or Hindustani or Swahili before an audience that understood only English, no one would blush or be angry; the story would be neither shocking nor obscene—indeed, it would not even be a story. Likewise, the value of a dollar bill is not in the bill, but in our social agreement to accept it as a symbol of value. If that agreement were to break down through the collapse of our Government, the dollar bill would become only a scrap of paper. We do not understand a dollar bill by staring at it long and hard. We understand it by observing how people act with respect to it. We understand it by understanding the social mechanisms and the loyalties that keep it meaningful. Semantics is therefore a social study, basic to all other social studies.

It is often remarked that words are tricky—and that we are all prone to be deceived by "fast talkers," such as high-pressure salesmen, skillful propagandists, politicians or lawyers. Since few of us are aware of the degree to which we use words to deceive ourselves, the sin of "using words in a tricky way" is one that is always attributed to the other fellow. When the Russians use the word "democracy" to mean something quite different from what we mean by it, we at once accuse them of "propaganda," or "corrupting the meanings of words." But when we use the word "democracy" in the United States to mean something quite different from what the Russians mean by it, they are equally quick to accuse us of "hypocrisy." We all tend to believe that the way we use words is the correct way, and that people who use the same words in other ways are either ignorant or dishonest.

WORDS EVOKE DIFFERENT RESPONSES

Leaving aside for a moment such abstract and difficult terms as "democracy," let us examine a common, everyday word like "frog." Surely there is no problem about what "frog" means! Here are some sample sentences:

"If we're going fishing, we'll have to catch some frogs first." (This is easy.)

"I have a frog in my throat." (You can hear it croaking.)

"She wore a loose, silk jacket fastened with braided frogs."

"The blacksmith pared down the frog and the hoof before shoeing the horse."

"In Hamilton, Ohio, there is a firm by the name of American Frog and Switch Company."

In addition to these "frogs," there is the frog in which a sword is carried, the frog at the bottom of a bowl or vase that is used in flower arrangement, and the frog which is part of a violin bow. The reader can no doubt think of other "frogs."

Or take another common word such as "order." There is the *order* that the salesman tries to get, which is quite different from the *order* which a captain gives to his crew. Some people enter holy *orders*. There is the *order* in the house when mother has finished tidying up; there is the batting *order* of the home team; there is an *order* of ham and eggs. It is surprising that with so many meanings to the word, people don't misunderstand one another oftener than they do.

The foregoing are only striking examples of a principle to which we are all so well accustomed that we rarely think of it; namely, that most words have more meanings than dictionaries can keep track of. And when we consider further that each of us has different experiences, different memories, different likes and dislikes, it is clear that all words evoke different responses in all of us. We may agree as to what the term "Mississippi River" stands for, but you and I recall different experiences with it; one of us has read more about it than the other; one of us may have happy memories of it, while the other may recall chiefly tragic events connected with it. Hence your "Mississippi River" can never be identical with my "Mississippi River." The fact that we can communicate with each other about the "Mississippi River" often conceals the fact that we are talking about two different sets of memories and experiences.

FIXED REACTIONS TO CERTAIN WORDS

Words being as varied in their meanings as they are, no one can tell us what the correct interpretation of a word should be in advance of our next encounter with that word. The reader may have been taught always to revere the word "mother." But what is he going to do the next time he encounters this word, when it occurs in the sentence "Mother began to form in the bottle"? If it is impossible to determine what a single word will mean on next encounter, is it possible to say in advance what is the correct evaluation of such events as these: (1) next summer, an individual who calls himself a socialist will announce his candidacy for the office of register of deeds in your city; (2) next autumn, there will be a strike at one of your local department stores; (3) next week, your wife will announce that she is going to change her style of hairdo; (4) tomorrow your little boy will come home with a bleeding nose?

A reasonably sane individual will react to each of these events in his own way, according to time, place and the entire surrounding set of circumstances; and included among those circumstances will be his own stock of experiences, wishes, hopes and fears. But there are people whose pattern of reactions is such that some of them can be completely predicted in advance. Mr. A will never vote for anyone called "socialist," no matter how incompetent or crooked the alternative candidates may be. Mr. B-1 always disapproves of strikes and strikers, without bothering to inquire whether or not this strike has its justifications; Mr. B-2 always sympathizes with the strikers because he hates all bosses. Mr. C belongs to the "stay as sweet as you are" school of thought, so that his wife hasn't been able to change her hairdo since she left high school. Mr. D always faints at the sight of blood.

Such fixed and unalterable patterns of reaction—in their more obvious forms we call them prejudices—are almost inevitably organized around words. Mr. E distrusts and fears all people to whom the term "Catholic" is applicable, while Mr. F, who is Catholic, distrusts and fears all non-Catholics. Mr. G is so rabid a Republican that he reacts with equal dislike to all Democrats, all Democratic proposals, all opposite proposals if they are also made by Democrats. Back in the days when Franklin D. Roosevelt was President, Mr. G disliked not only the Democratic President but also his wife, children and dog. His office was on Roosevelt Road in Chicago (it had been named after Theodore Roosevelt), but he had his address changed to his back door on 11th Street, so that he would not have to print the hated name on his stationery. Mr. H, on the other hand, is an equally rabid Democrat, who hates himself for continuing to play golf, since golf is Mr. Eisenhower's favorite game. People suffering from such prejudices seem to have in their brains an uninsulated spot which, when touched by such words as "capitalist," "boss," "striker," "scab," "Democrat," "Republican," "socialized medicine," and other such loaded terms, results in an immediate short circuit, often with a blowing of fuses.

Alfred Korzybski, the founder of general semantics, called such short-circuited responses "identification reactions." He used the word "identification" in a special sense; he meant that persons given to such fixed patterns of response identify (that is, treat as identical) all occurrences of a given word or symbol; they identify all the different cases that fall under the same name. Thus, if one has hostile identification reactions to "women drivers," then all women who drive cars are "identical" in their incompetence.

Korzybski believed that the term "identification reaction" could be generally used to describe the majority of cases of semantic malfunctioning. Identification is something that goes on in the human nervous system. "Out there" there are no absolute identities. No two Harvard men, no two Ford cars, no two mothers-in-law, no two politicians, no two leaves from the same tree, are identical with each other in all respects. If, however, we treat all cases that fall under the same class label as one at times when the differences are important, then there is something wrong with our semantic habits.

ANOTHER DEFINITION OF GENERAL SEMANTICS

We are now ready, then, for another definition of general semantics. It is a comparative study of the kinds of responses people make to the symbols and signs around them; we may compare the semantic habits common among the prejudiced, the foolish and the mentally ill with those found among people who are able to solve their problems successfully, so that, if we care to, we may revise our own semantic habits for the better. In other words, general semantics is, if we wish to make it so, the study of how not to be a damn fool.

Identification reactions run all the way through nature. The capacity for seeing similarities is necessary to the survival of all animals. The pickerel, I suppose, identifies all shiny, fluttery things going through the water as minnows, and goes after them all in pretty much the same way. Under natural conditions, life is made possible for the pickerel by this capacity. Once in a while, however, the shiny, fluttery thing in the water may happen to be not a minnow but an artificial lure on the end of a line. In such a case, one would say that the identification response, so useful for survival, under somewhat more complex conditions that require differentiation between two sorts of shiny and fluttery objects, proves to be fatal.

To go back to our discussion of human behavior, we see at once that the problem of adequate differentiation is immeasurably more complex for men than it is for pickerel. The signs we respond to, and the symbols we create and train ourselves to respond to, are infinitely greater in number and immeasurably more abstract than the signs in a pickerel's environment.

Lower animals have to deal only with certain brute facts in their physical environment. But think, only for a moment, of what constitutes a human environment. Think of the items that call for adequate responses that no animal ever has to think about: our days are named and numbered, so that we have birthdays, anniversaries, holidays, centennials, and so on, all calling for specifically human responses: we have history, which no animal has to worry about; we have verbally codified patterns of behavior which we call law, religion and ethics. We have to respond not only to events in our immediate environment, but to reported events in Washington, Paris, Tokyo, Moscow, Beirut. We have literature, comic strips, confession magazines, market quotations, detective stories, journals of abnormal psychology, bookkeeping systems to interpret. We have money, credit, banking, stocks, bonds, checks, bills. We have the complex symbolisms of moving pictures, paintings, drama, music, architecture and dress. In short, we live in a vast human dimension of which the lower animals have no inkling and we have to have a capacity for differentiation adequate to the complexity of our extra environment.

WHY DO PEOPLE REACT AS THEY DO?

The next question, then, is why human beings do not always have an adequate capacity for differentiation. Why are we not constantly on the lookout for differences as well as similarities instead of feeling, as so many do, that the Chinese (or Mexicans, or ballplayers, or women drivers) are "all alike"? Why do some people react to words as if they were the things they stand for? Why do certain patterns of reaction, both in individuals and in larger groups such as nations, persist long after the usefulness has expired?

Part of our identification reactions are simply protective mechanisms inherited from the necessities of survival under earlier and more primitive conditions of life. I was once beaten up and robbed by two men on a dark street. Months later, I was again on a dark street with two men, good friends of mine, but involuntarily I found myself in a panic and insisted on our hurrying to a well-lighted drugstore to have a soda so that I would stop being jittery. In other words, my whole body reacted with an identification reaction of fear of these two men, in spite of the fact that "I knew" that I was in no danger. Fortunately, with the passage of time, this reaction has died away. But the hurtful experiences of early childhood do not fade so readily. There is no doubt that many identification reactions are traceable to childhood traumas, as psychiatrists have shown.

Further identification reactions are caused by communal patterns of behavior which were necessary or thought necessary at one stage or another in the development of a tribe or nation. General directives such as "Kill all snakes," "Never kill cows, which are sacred animals," "Shoot all strangers on sight," "Fall down flat on your face before all members of the aristocracy," or, to come to more modern instances, "Never vote for a Republican," "Oppose all government regulation of business," "Never associate with Negroes on terms of equality," are an enormous factor in the creation of identification reactions.

Some human beings—possibly in their private feelings a majority—can accept these directives in a *human* way: that is, it will not be impossible for them under a sufficiently changed set of circumstances to kill a cow, or not to bow down before an aristocrat, to vote for a Republican, or to accept a Negro as a classmate. Others, however, get these directives so deeply ground into their nervous systems that they become incapable of changing their responses no matter how greatly the circumstances may have changed. Still others, although capable of changing their responses, dare not do so for fear of public opinion. Social progress usually requires the breaking up of these absolute identifications, which often makes necessary changes impossible. Society must obviously have patterns of behavior; human beings must obviously have habits. But when those patterns become inflexible, so that a tribe has only one way to meet a famine, namely, to throw more infants as sacrifices to the crocodiles, or a nation has only one way to meet a threat to its security, namely, to increase its armaments, then such a tribe or such a nation is headed for trouble. There is insufficient capacity for differentiated behavior.

Furthermore—and here one must touch upon the role of newspapers, radio and television—if agencies of mass communication hammer away incessantly at the production of, let us say, a hostile set of reactions to such words as "Communists," "bureaucrats," "Wall Street," "international bankers," "labor leaders," and so on, no matter how useful an immediate job they may perform in correcting a given abuse at a given time and place, they can in the long run produce in thousands of readers and listeners identification reactions to the words—reactions that will make intelligent public discussion impossible. Modern means of mass communication and propaganda certainly have an important part to play in the creation of identification reactions.

In addition to the foregoing, there is still another source of identification reactions; namely, the language we use in our daily thought and speech. Unlike the languages of the sciences, which are carefully constructed, tailor-made, special-purpose languages, the language of everyday life is one directly inherited and haphazardly developed from those of our prescientific ancestors: Anglo-Saxons, primitive Germanic tribes, primitive Indo-Europeans. With their scant knowledge of the world, they formulated descriptions of the world before them in statements such as "The sun rises." We do not today believe that the sun "rises." Nevertheless, we still continue to use the expression, without believing what we say.

ERRONEOUS IMPLICATIONS IN EVERYDAY LANGUAGE

But there are other expressions, quite as primitive as the idea of "sunrise," which we use uncritically, fully believing in the implications of our terms. Having observed (or heard) that *some* Negroes are lazy, an individual may say, making a huge jump beyond the known facts, "Negroes are lazy." Without arguing for the moment the truth or falsity of this statement, let us examine the implications of the statement as it is ordinarily constructed: "Negroes are lazy." The statement implies, as common sense or any textbook on traditional logic will tell us, that "laziness" is a "quality" that is "inherent" in Negroes.

What are the facts? Under conditions of slavery, under which Negroes were not paid for working, there wasn't any point in being an industrious and responsible worker. The distinguished French abstract artist Jean Hélion once told the story of his life as a prisoner of war in a German camp, where, during the Second World War, he was compelled to do forced labor. He told how he loafed on the job, how he thought of device after device for avoiding work and producing as little as possible—and, since his prison camp was a farm, how he stole chickens at every opportunity. He also described how he put on an expression of good-natured imbecility whenever approached by his Nazi overseers. Without intending to do so, in describing his own actions, he gave an almost perfect picture of the literary type of the Southern Negro of slavery days. Jean Hélion, confronted with the fact of forced labor, reacted as intelligently as Southern Negro slaves, and the slaves reacted as intelligently as Jean Hélion. "Laziness," then, is not an "inherent quality" of Negroes or of any other group of people. It is a *response* to a work situation in which there are no rewards for working, and in which one hates his taskmasters.

Statements implying inherent qualities, such as "Negroes are lazy" or "There's something terribly wrong with young people today," are therefore the crudest kind of unscientific observation, based on an out-of-date way of saying things, like "The sun rises." The tragedy is not simply the fact that people make such statements; the graver fact is that they believe themselves.

Some individuals are admired for their "realism" because, as the saying goes, they "call a spade a spade." Suppose we were to raise the question "Why should anyone call it a spade?" The reply would obviously be, "Because that's what it is!" This reply appeals so strongly to the common sense of most people that they feel that at this point discussion can be closed. I should like to ask the reader, however, to consider a point which may appear at first to him a mere quibble.

Here, let us say, is an implement for digging made of steel, with a wooden handle. Here, on the other hand, is a succession of sounds made with the tongue, lips and vocal cords: "spade." If you want a digging implement of the kind we are talking about, you would ask for it by making the succession of sounds "spade" if you are addressing an English-speaking person. But suppose you were addressing a speaker of Dutch, French, Hungarian, Chinese, Tagalog? Would you not have to make completely different sounds? It is apparent, then, that the common-sense opinion of most people, "We call a spade a spade because that's what it is," is completely and utterly wrong. We call it a "spade" because we are English-speaking people, conforming, in this instance, to majority usage in naming this particular object. The steel-and-iron digging implement is simply an object standing there against the garage door; "spade" is what we *call* it—"spade" is a *name*.

And here we come to another source of identification reactions—an unconscious assumption about language epitomized in the expression "a spade is a spade," or even more elegantly in the famous remark "Pigs are called pigs because they are such dirty animals." The assumption is that everything has a "right name" and that the "right name" names the "essence" of that which is named.

If this assumption is at work in our reaction patterns, we are likely to be given to premature and often extremely inappropriate responses. We are likely to react to names as if they gave complete insight into the persons, things or situations named. If we are told that a given individual is a "Jew," some of us are likely to respond, "That's all I need to know." For, if names give the essence of that which is named, obviously, every "Jew" has the essential attribute of "Jewishness." Or, to put it the other way around, it is because he possesses "Jewishness" that we call him a "Jew"! A further example of the operation of this assumption is that, in spite of the fact that my entire education has been in Canada and the United States and I am unable to read and write Japanese, I am sometimes credited, or accused, of having an "Oriental mind." Now, since Buddha, Confucius, General Tojo, Mao Tse-tung, Syngman Rhee, Pandit Nehru and the proprietor of the Golden Pheasant Chop Suey House all have "Oriental minds," it is hard to imagine what is meant. The "Oriental mind," like the attribute of "Jewishness," is purely and simply a fiction. Nevertheless, I used to note with alarm that newspaper columnists got paid for articles that purported to account for Stalin's behavior by pointing out that since he came from Georgia, which is next to Turkey and Azerbaijan and therefore "more a part of Asia than of Europe," he too had an "Oriental mind."

IMPROVING YOUR SEMANTIC HABITS

Our everyday habits of speech and our unconscious assumptions about the relations between words and things lead, then, to an identification reaction in which it is felt that all things that have the same name are entitled to the same response. From this point of view, all "insurance men," or "college boys," or "politicians," or "lawyers," or "Texans" are alike. Once we recognize the absurdity of these identification reactions based on identities of name, we can begin to think more clearly and more adequately. No "Texan" is exactly like any other "Texan." No "college boy" is exactly like any other "college boy." Most of the time "Texans" or "college boys" may be what you think they are: but often they are not. To realize fully the difference between words and what they stand for is to be ready for differences as well as similarities in the world. This readiness is mandatory to scientific thinking, as well as to sane thinking.

Korzybski's simple but powerful suggestion to those wishing to improve their semantic habits is to add "index numbers" to all terms, according to the formula: A_1 is not A_2. Translated into everyday language we can state the formula in such terms as these: Cow_1 is not cow_2; cow_2 is not cow_3; $Texan_1$ is not $Texan_2$; $politician_1$ is not $politician_2$; ham and eggs (Plaza Hotel) are not ham and eggs (Smitty's Café); socialism (Russia) is not socialism (England); private enterprise (Joe's Shoe Repair Shop) is not private enterprise (A.T.&T.). The formula means that instead of simply thinking about "cows" or "politicians" or "private enterprise," we should think as factually as possible about the differences between one cow and another, one politician and another, one privately owned enterprise and another.

The device of "indexing" will not automatically make us wiser and better, but it's a start. When we talk or write, the habit of indexing our general terms will reduce our tendency to wild and woolly generalization. It will compel us to think before we speak—think in terms of concrete objects and events and situations, rather than in terms of verbal associations. When we read or listen, the habit of indexing will help us visualize more concretely, and therefore understand better, what is being said. And if nothing is being said except deceptive windbaggery, the habit of indexing may—at least part of the time—save us from snapping, like the pickerel, at phony minnows. Another way of summing up is to remember, as Wendell Johnson said, that "To a mouse, cheese is cheese—that's why mousetraps work."

DISCUSSION AND ACTIVITY

1. What are some of the euphemisms, or synonyms with dissembling connotations, for words mentioned by Hayakawa like "Communists" and "bureaucrats"? Are the identification reactions the same for the euphemisms as for the words they stand for? Why do some words cause us to react negatively while others with essentially the same meanings cause us to react neutrally or favorably?

2. Why should writers in business and industry pay special attention to the words they use?

Suggestions for Further Reading

Adelstein, Michael E. "Style." *Contemporary Business Writing.* Random House, New York, 1971.

Style is determined not only by a writer's words, grammar, brevity, vigor, and clarity, but also by a writer's attitude, tone, sincerity, and point of view. An effective style is one characterized by a tone that is warm, friendly, natural, and courteous; a sincerity brought about by using details, "people" nouns, first-person pronouns, and a *you*-attitude; and an appropriate point of view, whether personal or impersonal.

Bromage, Mary C. "Gamemanship in Written Communication." *Management Review* 61 (April 1972), 11–15.

Fearing reprisals for their judgments, opinions, or new ideas, some writers in business and industry deliberately obfuscate their messages by resorting to stylistic devices that impede clarity. Among the devices used to obscure meanings are abstract words, subjective vocabulary, passive voice, bland language, clichés, jargon, and long sentences.

Cousins, Norman. "Completing the Scholar's Education." *Saturday Review,* September 2, 1950, pp. 22–23.

Especially among academic writers, "there is fear of simplicity lest one be thought simple; fear of clarity lest one be thought transparent; fear of lucidity lest one be thought glib. . . . [Yet] transmission of ideas calls for the organization of logic, set down in sequence, precision, and with maximum clarity."

Dandridge, Edmund P., Jr. "Notes Toward a Definition of Technical Writing." *Journal of Technical Writing and Communication* 3 (1973), 265–271.

An analysis of ten samples of good technical writing and of ten samples of good nontechnical writing supports the assumption that good technical writing is generally stylistically simpler than good nontechnical writing. When contrasted with the nontechnical samples, the technical samples tend to use fewer words, shorter sentences, simpler sentences, more paragraphs, and shorter paragraphs.

Davis, Richard M. "The Way We Write It Makes a Difference—Sometimes: An Experiment in the Effects of Non-Standard Expression." *Journal of Business Communication* 6 (1968), 15–22.

The way one writes is at times not as important as one assumes. Departures from standard English do not always interfere with the clear transmission of messages.

Gregg, Marjorie True. "Cobblestone Style." *Saturday Review*, June 4, 1932, p. 771.

American writers use too many noun constructions and not enough action verbs. "Cut out the noun constructions that are clogging and clotting and curdling your language. . . . *Use active verbs.*"

Orwell, George. "Politics and the English Language." *Shooting an Elephant and Other Essays.* Harcourt Brace, New York, 1950.

Stale, imprecise language often befuddles modern readers. To avoid such a bromidic style in one's own writing, one should rely on these six rules: (1) Never use a metaphor, simile, or other figure of speech that one is used to seeing in print. (2) Never use a long word where a short one will do. (3) If it is possible to cut a word out, always cut it out. (4) Never use the passive voice where one can use the active. (5) Never use a foreign phrase, a scientific word, or a jargon word if one can think of an everyday English equivalent. (6) One should break any of these rules rather than writing anything "outright barbarous."

Shepley, James R. "A Letter from the Publisher." *Time*, September 20, 1968, p. 13.

An objective style is impossible to achieve in reporting. However, fairness can be achieved through an interplay of reliable facts and balanced judgment.

Stevenson, Dwight W., and J. C. Mathes. "Technical Writing: New Name or New Course?" *Teaching English in the Two-Year College* 2 (1975), 15–19.

Stylistic differences exist between technical exposition and other kinds of exposition. For example, variety and novelty in sentence patterns, which are virtues in ordinary prose, might result in technical prose whose clarity and effectiveness are lost.

Thurber, James. "Which." *Effective English*, edited by George W. Crouch. American Institute of Banking, New York, 1959.

When used recklessly, the relative pronoun "which" often proves troublesome to writers and readers. One should remember that "one 'which' leads to two and that two 'whiches' multiply like rabbits."

Part III

What Are Some Important
Writing Strategies?

*"True ease in writing comes from art, not chance,
As those move easiest who have learned to dance."*
Alexander Pope*

Your success as a writer depends largely on your ability to choose and execute the writing strategy that does the job best. "Strategy" comes from the Greek word *strategia* meaning "generalship" or "office of the general." Military strategy is the art of the commander to plan and coordinate the overall conduct of combat operations. When you are up against a writing situation, you must become a writing strategist—that is, you must know how to marshall your resources, plan your writing, and achieve your goal. That goal is always a well-written, attractively designed piece of writing that engages your reader. It is a goal that is achieved by conscious strategy, not by accident. Just how conscious your strategy must be is explained in the articles in this section.

As a writing strategist facing various writing situations, you need to know how to solve many problems: what is the most efficient way to write, what are effective ways to engage your reader, what are the standard ways to organize information, what is the best format for your report? The first two problems are discussed in this section. The last two are discussed in Parts IV and V.

There are, of course, many ways to go about writing a letter or report. But if you learn a standard procedure for writing, your efficiency and output will increase. A standard procedure develops out of a myriad of attempts to do something. Through experience and through refinement by

* From "An Essay on Criticism."

experts, a particular version of the procedure is made "standard" and the task is somewhat simplified. Ultimately, the standard procedure becomes the acceptable way of doing things. It is sort of like a Darwinian survival of the fittest. Being familiar with the standard procedure does not automatically guarantee that you can perform the work perfectly. But once you know the procedure to follow, practice and use will make your work more efficient and predictable.

Two such standardized procedures are explained in the first two articles in this section. Michael Adelstein's "Writing Is Work" divides the writing process into five major stages. Notice the name of each stage, study what each stage involves, and remember the amount of time allocated to each stage. Perhaps you will be surprised at the time devoted to revising. Revising is the key to shaping your writing to achieve your goals, and you court disaster if you do not revise. L. L. Farkas's "Writing Better Technical Papers" describes a standard procedure for writing papers from inception to final editing. You can apply the procedure to all kinds of writing.

As a well-known axiom goes: "audience comes first." And this notion, perhaps more than any other, distinguishes business and technical writing from other types of writing. In other writing—most notably the "creative" writing of fiction and poetry—readers come in a distant third to writers and their subjects. Depending on what literary theory you accept, primary importance is placed either on the special character of novelists and poets or on the worlds created by them. For the most part, readers are assumed to be ideal readers, fully prepared to relate to the fiction and poetry on the author's terms. This expectation is as it should be; it is appropriate for what we regard as creative writing.

But a different expectation exists in business and technical writing, where readers are busy executives who want the important findings up front, or are privates last-class who need information at a level they can understand, or somewhere in the bewildering range between. Racker's "Selecting and Writing to the Proper Level" identifies the major levels of audiences and explains how to reach them. H. R. Clauser's "Writing Is for Readers" stresses that understanding how your audience has been trained to read will help you handle details. Adapting material to your audience is almost an all-or-nothing situation in which relevancy and clarity—from your reader's point of view—are the keys to successful communication.

Two other matters concerned with engaging your reader, and closely related to audience adaptation, are the use of graphics and the use of quotations from the writings of others.

Learning to use graphics will enlarge your communication skills greatly. In *The Art of Readable Writing*, Rudolf Flesch asks a surprising question: "Are words necessary?" The answer of course is that words are indispensable. However, graphics make mathematical concepts easier to understand, represent details or relationships of mechanisms and processes with much greater impact than can words alone, condense and highlight material, and arouse reader interest.

Some graphics (photographs and drawings) help readers visualize objects. Words can go only so far in helping show what things look like. On page 203, Allen provides an example of the Job Analysis Work Sheet (reduced size) to show readers exactly what the worksheet looks like. In "How to Type Your Paper," Shelby includes eight graphics to show the layout requirements of formal elements of reports. If either writer had relied solely on words to explain, it would be difficult, if not impossible, to visualize these documents. DeBakey's cartoons accompanying "The Persuasive Proposal" clarify the concepts she is presenting by making visual the images behind language and sustaining reader interest.

Other graphics (tables and charts) highlight important information for readers. Nearly every selection in this book uses lists, either informally inserted in paragraphs or formally boxed and labeled with a table number and title. At the end of this section, following the selections, are summaries of five articles on graphics that will be of immense help to you.

The skillful handling of quoted material in your reports helps your reader follow *your* writing. Perhaps nothing will challenge you more than orchestrating someone else's knowledge in your own presentation. If quotations are handled ineptly, you lose the sense of your own voice in writing and lose control of your report. You must avoid merely dumping quoted material into your report. You must skillfully mix the knowledge of others with your own writing in order to illustrate and support your own thoughts. Otherwise, your reader will get lost among your references and will not be able to tell who is saying what. As Jacques Barzun and Henry F. Graff point out in their article, quoting is a very sophisticated art.

Selection Fourteen

Writing Is Work

MICHAEL E. ADELSTEIN

Michael E. Adelstein, professor of English and director of freshman English at the University of Kentucky, has taught business writing at the university and in industry. Adelstein examines three misconceptions about the nature of writing and describes the process of writing.

Students often fail to realize that many of the skills required in business can be learned from fellow employees, in-service training courses, or personal study. But the art of writing, like that of piano playing or painting, can be perfected only with instruction, practice, and supervision. And while instruction is easy to obtain from books like this one, and practice requires only determination, supervision is difficult to find. Many people can recognize and identify poor writing; few besides experienced English teachers can point out why it is poor and what can be done to improve it. Consequently, if college graduates want to write well, they will be wise to learn to do so before leaving the campus.

If writing well is important in business, and if mastery of this art is such a valuable asset, why do students balk at taking writing courses? Let me suggest three reasons: (1) few students realize the amount and significance of writing in business; (2) few students enjoy writing; and (3) few students are willing to labor strenuously to achieve the perfection good writing demands.

Most people picture the business executive as a man of action, pounding his desk, exhorting salesmen, dashing for a plane, phoning from his car, landing a lucrative contract, and closing a big business deal. Neither television nor the movies show him opening his attaché case at home or in the office, sitting down at his desk, and struggling, sometimes for hours, to draft a report or proposal. Nor do they show the time and energy that he devotes every day to writing.

But even if the students realized how much writing they will have to do in business, and how valuable the ability to write well will prove to be, they would still seek to avoid writing courses. The reason is natural. Most of us enjoy activities we do well and dislike those we do poorly. The more we excel at golf, water-skiing, or tennis, the greater the pleasure we derive from these sports. But if we are inept at something—perhaps bowling, or ice-skating, or bridge—then it is no fun to participate. So it is with writing. Many male students were bored in high school English courses, attained little proficiency in using language, and barely survived freshman English. Failing to realize the importance of writing but realizing their own failure, these students naturally find writing tedious.

But few people have ever claimed that writing is fun. Dr. Samuel Johnson, the eighteenth-century monarch of letters, put it savagely and succinctly, "No man but a blockhead ever wrote for anything but money." Many others testify to the excruciating agony of writing. Much of the frustration stems from the fact that the more a person learns about writing, the more difficult it usually becomes. If a store has only one suit in a customer's correct size, then his decision is simple. But if it has many, then the problem is more complex. So with writing. When a person becomes aware of the almost infinite number of words and phrases to choose from and of the innumerable ways to arrange them in sentences and paragraphs to convey an idea clearly and directly, then his selection requires study, analyzing, evaluating, and determining, all forms of the hardest and highest human endeavor—thinking.

At this point you might well ask, "If writing becomes more difficult as I learn more about it, why study it?" The answer: to improve. And you will improve—but only if you are willing to work hard. You will acquire a skill that may be more valuable than anything else you will learn in college. But the going is difficult, the progress slow, and the end never completely attained. To write well necessitates the same kind of fortitude, concentration, and painstaking precision needed to master a complicated cost-accounting problem or to design an intricate computer program. Get rid of any ideas that writing is fun. See it for what it is—work! Ernest Hemingway states it well: "What you ultimately remember about anything you've written is how difficult it was to write it."

Like all marketable skills and commodities, writing ability is affected by supply and demand. The demand is almost insatiable: hundreds of thousands of people are this very moment busy at writing articles, memos, letters, reports, pamphlets, public-relations stories, bulletins, circulars, advertising copy, and instructions. The supply, however, is limited: few people write well. Therefore, good writers are highly valuable in any organization.

It would seem that more people would have learned to write. But they failed to acquire this valuable skill because they were unaware in school how important it was, and because they labored under several delusions. These delusions might be called the myths of genius, inspiration, and correctness.

THE MYTH OF GENIUS

Some people feel that writers are born, not made. These individuals contend that only a few possess the necessary genius to write well. There undoubtedly is some truth to this view. Not many people have the imagination, sensitivity, insight, or talent to become fine essayists, poets, or novelists. But nearly anyone can learn to express his ideas clearly and convincingly.

Of course, it may take some people a great deal of time and effort to acquire this skill, just as some people take longer than others to learn how to drive a car. But just as most people eventually pass a driver's test, so most people can learn to write proficiently. Naturally, writing demands more intellectual effort than driving, requires more training, and is affected more by a person's environment, education, and intelligence. But anyone in college can learn to write competently if he desires to do so and if he receives proper instruction, practice, and supervision.

The beauty and grace of fine prose may be beyond the reach of most people, but the clarity and accuracy of everyday functional prose are within their grasp. But many people prefer believing that they lack the necessary genius, rather than working hard to master the principles of writing.

THE MYTH OF INSPIRATION

Then there are people who contend that they have to be inspired to write well. This species, common in high school and college, consists of students who fuss and fret about assignments, claiming that they must be excited about a subject before they can do justice to it. Nonsense! We are seldom provoked enough to sit down to write, as the small flow of letters to the national television networks can attest. And few of us are ever inspired to write. Of course, some poets—like Wordsworth and Coleridge—waited for inspiration and were rewarded. But most people wait in vain. Professional writers, placing their faith in perspiration, force themselves to write on schedule, day in and day out, whether they want to or not. Sometimes, like J. D. Salinger, author of *Catcher in the Rye*, they imprison themselves in isolated places, removed from the disconcerting distraction of telephones, television, radio, refrigerator, and doorbells. Sometimes, practically following the advice of Anthony Trollope, a nineteenth-century novelist who suggested that writers glue themselves to chairs, they remain dutifully at their desks eight hours a day, even if they cannot compose a single sentence.

The truth about writing is that it rarely results from inspiration and seldom comes easily. And, for several reasons, it is even more difficult in business than in school. Usually a teacher will assign several subjects to provide students with at least one interesting topic. But in business, people have little choice, being frequently required to write about matters they know little of and care for less. In addition, everything usually must be done in a hurry. The letter, memo, or report should have been sent yesterday; therefore, it must be written immediately. Under such circumstances, a person cannot dally for inspiration. And if he did, it probably would not come.

THE MYTH OF CORRECTNESS

Most people believe that grammar is all one needs to know to write well. I can vividly recall teaching Business English to executives from IBM and other corporations who insisted that what they wanted was to be drilled in grammar and to complete exercises in workbooks. I can also recall hundreds of students who were so nervous about grammar, spelling, punctuation, and capitalization that they wrote lifeless, petrified prose. What these businessmen, students, and numerous others like them need to realize is that correctness is only one aspect of good writing, and that for most people, it is a minor one. Raised in an era of mass media, most of us have been conditioned to what is proper as a result of our constant reading of newspapers and magazines, and our incessant listening to radio and television. Some sticky usage problems still do plague us, but, in general, people would greatly improve their writing if they could get over their stage fright about grammar.

THE WRITING PROCESS

Their prevalent belief in the myths of genius, inspiration, and correctness prevents many people from writing effectively. If these individuals could realize that they can learn to write, that they cannot wait to be inspired, and that they must focus on other aspects besides correctness, then they can overcome many obstacles confronting them. But they should be aware that writing, like many other forms of work, is a process. To accomplish it well, people should plan on completing each one of its five stages. The time allocations may vary with the deadline, subject, and purpose, but the work schedule should generally follow this pattern:

1. Worrying—15 percent
2. Planning—10 percent
3. Writing—25 percent
4. Revising—45 percent
5. Proofreading—5 percent

Note that only 25 percent of the time should be spent in writing; the rest—75 percent—should be spent in getting ready for the task and perfecting the initial effort. Observe also that more time is spent in revising than in any other stage, including writing.

STAGE ONE—WORRYING (15 PERCENT)

Worrying is a more appropriate term for the first operation than *thinking* because it suggests more precisely what you must do. When you receive a writing assignment, you must avoid blocking it out of your mind until you're ready to write; instead, allow it to simmer while you try to cook up some ideas. As you stew about the subject while brushing your teeth, taking a shower, getting dressed, walking to class, or preparing for bed, jot down any pertinent thoughts. You need only note them on scrap paper or the back of an envelope. But if you fail to write them down, you'll forget them.

Another way to relieve your worrying is to read about the subject or discuss it with friends. The chances are that others have wrestled with similar problems and written their ideas. Why not benefit from their experience? But if you think that there is nothing in print about the subject or a related one—either because your assignment is restricted, localized, or topical—then you should at least discuss it with friends.

Let's say, for example, that as a member of the student government on your campus, you've been asked to investigate the possibility of opening a student-operated bookstore to combat high textbook prices. During free hours, you might drop into the library to read about the retail book business and, if possible, about college bookstores. If your student government is affiliated with the National Student Association, you might write to it for information about policies and practices on other campuses. In addition, you might start talking about the problem to the manager of your college bookstore, owners of other bookstores in your community, student transfers from other schools, managers of college bookstores in nearby communities, and some of your professors. Because you might be concerned about the public-relations repercussions of a student bookstore, you should propose the venture to college officials. At this point, you will have heard many different ideas. To discuss some of them, bring up the subject of a student bookstore with friends. As you talk, you will be clarifying and organizing your thoughts.

The same technique can be applied to less involved subjects, such as a description of an ideal summer job, or an explanation of a technical matter like input-output analysis. By worrying about the subject as you proceed through your daily routine, by striving to get ideas from books and people or both, and by clarifying your thoughts in conversation with friends, you will be ready for the next step. Remember: you can proceed only if you have some ideas. You must have something to write about.

STAGE TWO—PLANNING (10 PERCENT)

Planning is another term for organizing or for the task that students dread so—*outlining*. You should already have had some experience with this process. Like many students, unfortunately, you may have learned only how to outline a paper after writing it. This practice is a waste of time. If you want to operate efficiently, then it follows that you will have to specialize in each of the writing stages in turn. When you force yourself simultaneously to conceive of ideas, to arrange them in logical order, and to express them in words, you cannot perform all of these complex activities as effectively as if you had concentrated on each one separately. By dividing the writing process into five stages, and by focusing on each one individually, the odds are that the result will be much better than you had thought possible. Of course you will hear about people who never plan their writing but who do well anyway, just as you may know someone who skis or plays golf well without having had a lesson. There are always some naturals who can flaunt the rules. But perhaps these people could have significantly improved their writing, golf, or skiing if they had received some formal instruction and proceeded in the prescribed manner.

An efficient person plans his work. Many executives take a few minutes before leaving the office or retiring at night to jot down problems to attend to the following day. Many vacationers list tasks to do before leaving home: stop the newspaper, turn down the refrigerator, inform the mailman, cut off the hot water, check the car, mail the mortgage payment, notify the police, and the like. Of course, you can go on vacation without planning—just as you can write without planning—but the chances are that the more carefully you organize, the better the results will be.

In writing, planning consists mainly in examining all your ideas, eliminating the irrelevant ones, and arranging the others in a clear, logical order. Whether you write a formal outline or merely jot down your ideas on scrap paper is up to you. The outline is for your benefit; follow the procedure that helps you the most. But realize that the harder you work on perfecting your outline, the easier the writing will be. As you write, therefore, you can concentrate on formulating sentences instead of also being concerned with thinking of ideas and trying to organize them. For the present, realize the importance of planning. A later chapter will suggest some standard organizational practices and patterns to help you.

STAGE THREE—WRITING (25 PERCENT)

When you know what you want to say and have planned how to say it, then you are ready to write. This third step in the writing process is self-explanatory. You need only dash away your thoughts, using pen, pencil, or typewriter, whichever you prefer. Let your mind flow along the outline. Don't pause to check spelling, worry about punctuation, or search for exact words. If new thoughts pop into your mind, as they often do, check them with your outline, and work them into your paper if they are relevant. Otherwise discard them. Keep going until either the end or fatigue halts you. And don't worry if you write slowly: many people do.

STAGE FOUR—REVISING (45 PERCENT)

Few authors are so talented that they can express themselves clearly and effectively in a first draft. Most know that they must revise their papers extensively. So they roll up their sleeves and begin slashing away, cutting out excess verbiage, tearing into fuzzy sentences, stabbing at structural weaknesses, and knifing into obfuscation. Revision is painful: removing pet phrases and savory sentences is like getting rid of cherished possessions. Just as we dislike to discard old magazines, books, shoes, and clothes, so we hate to get rid of phrases and sentences. But no matter how distasteful the process may be, professional writers know that revising is crucial; it is usually the difference between a mediocre paper and an excellent one.

Few students revise their work well; they fail to realize the importance of this process, lack the necessary zeal, and are unaware of how to proceed. Once you understand why you seem to be naturally adverse to revision, you may overcome your resistance.

Because writing requires concentrated thought, it is about the most enervating work that humans perform. Consequently, upon completing a first draft, we are so relieved at having produced something that we tidy it up slightly, copy it quickly, and get rid of it immediately. After all, why bother with it anymore? But there's the rub. The experienced writer knows that he has to sweat through it again and again, looking for trouble, willing to rework cumbersome passages, and striving always to find a better word, a more felicitous phrase, and a smoother sentence. John Galbraith, for example, regularly writes five drafts, the last one reserved for inserting "that note of spontaneity" that ironically makes his work appear natural and effortless. The record for revision is probably held by Ernest Hemingway, who wrote the last page of *A Farewell to Arms* thirty-nine times before he was satisfied!

To revise effectively, you must not only change your attitude, but also your perspective. Usually we reread a paper from our own viewpoint, feeling proud of our accomplishment. We admire our cleverness, enjoy our graceful flights, and glow at our fanciful turns. How easy it is to delude ourselves! The experienced writer reads his draft objectively, looking at it from the standpoint of his reader. To accomplish this, he gets away from the paper for a while, usually leaving it until the following morning. You may not be able to budget your time this ideally, but you can put the paper aside while you visit a friend, grab a bite to eat, or phone someone. Unless you divorce yourself from the paper, you will probably be under its spell; that is, you will reread it with your mind rather than with your eyes. You will see only what you think is on the page instead of what is actually there. And you will not be able to transport yourself from your role of writer to that of critic.

Only by attacking your paper from the viewpoint of another person can you revise it effectively. You must be anxious to find fault and you must be honest with yourself. If you cannot or will not realize the weaknesses in your writing, then you cannot correct them. This textbook should open your eyes to many things that can go wrong, but above all you must *want* to find them. If you blind yourself to the bad, then your work will never be good. You must convince yourself that good writing depends on good rewriting. This point cannot be repeated too frequently or emphasized too much. Tolstoy revised *War and Peace*—one of the world's longest and greatest novels—five times. The brilliant French stylist Flaubert struggled for hours, even days, trying to perfect a single sentence. But question your professors about their own writing. Many—like me—revise papers three, four, or five times before submitting them for publication. You may not have this opportunity, your deadlines in school and business may not permit this luxury; but you can develop a respect for revision and you can devote more time and effort to this crucial stage in the writing process.

STAGE FIVE—PROOFREADING (5 PERCENT)

Worrying, planning, writing, and revising are not sufficient. The final task, although not as taxing as the others, is just as vital. Proofreading is like a quick check in the mirror before leaving for a date. A sloppy appearance can spoil a favorable acceptance of you or your paper. If, through your fault or your typist's, words are missing or repeated, letters are transposed, or sentences juxtaposed, a reader may be perturbed enough to ignore or resist your ideas or information. Poor proofreading of an application letter can cost you a job. Such penalties may seem unfair, but we all react in this fashion. What would you think of a doctor with dirty hands? His carelessness would not only disgust you but would also raise questions about his professional competence. Similarly, carelessness in your writing antagonizes readers and raises questions in their minds about your competence. The merit of a person or paper should not depend on appearances, but frequently it does. Being aware of this possibility, you should keep yourself and your writing well groomed.

Poor proofreading results from failure to realize its importance, from inadequate time, and from improper effort. Unless you are convinced that scrutinizing your final copy is important, you cannot proofread effectively. If you realize the significance of this fifth step in the writing process, then you will not only have the incentive to work hard at proofreading but also will allow time for it in planning.

Like most things, proofreading takes time. We usually run out of time for things that we don't care about. But there's always time for what we enjoy or what is important to us. So with proofreading; lack of time is seldom a legitimate excuse for doing it badly.

But lack of technique is. To proofread well, you must forget the ideas in a paper and focus on the words. Slow down your reading speed, stare hard at the black print, and search for trouble rather than trying to finish quickly (see Figure 14-1). Some professional proofreaders start with the last sentence and read backward to the first. Whatever technique you adopt, work painstakingly, so that a few careless errors will not spoil your efforts. If you realize the importance of proofreading, view it as one of the stages in the writing process, and labor at it conscientiously, your paper will reflect your care and concern.

EACH STAGE IS IMPORTANT

These then are the five stages in the writing process: worrying, planning, writing, revising, and proofreading. Each is vital. You need to become proficient at each of them, although, at some later point in your life, an editor or efficient secretary may relieve you of some proofreading chores. But in your career as a student and in your early business years, you will be on your own to guide your paper through the five stages. Since it is your paper, no one else will be as interested in it as you. It's your offspring—to nourish, to cherish, to coddle, to bring up, and finally, to turn over to someone else. You will be proud of it only if you have done your best throughout its growth and development.

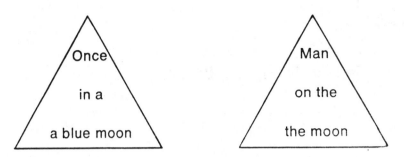

Figure 14-1. A simple proofreading exercise. Simple?

DISCUSSION AND ACTIVITY

1. Why do students balk at taking writing courses?

2. What are the three delusions based upon myths about writing that hinder people in learning how to write?

3. Now that you have read Adelstein's description of the writing process, relate it to your own past writing practices. How closely does your practice conform to the steps and percentages of time devoted to each step as described by Adelstein? Are you surprised at the various percentages spent in the steps? Do you leave out any steps in your own practice?

4. If your writing practice differs from that described by Adelstein, are you going to use the process described by Adelstein the next time you write? If not, why not?

5. Conduct a survey of ten students at your school (but not enrolled in your business or technical writing course), asking them to identify the steps in the writing process. Or conduct a survey of ten students asking them to respond to Adelstein's five steps by saying what they think is the percentage of time that should be spent on each step. Write a brief report that explains (1) what information you were seeking in the survey; (2) what method you used to get the information; (3) what the results were; and (4) what your conclusions are.

Selection Fifteen

Writing Better Technical Papers

L. L. FARKAS

Formerly the manager of Advanced and Special Programs for the Quality Division of Martin-Marietta Company, Orlando, Florida, L. L. Farkas describes a procedure for writing papers from beginning to final editing.

How do you go about writing a better technical paper for a magazine or a symposium? Where do you start? How do you organize it, write it? What does it need to be acceptable? To answer these questions let us discuss the details of writing such a paper from its inception to submittal.

ESTABLISH THE NEED

Before you put a single word on paper consider the subject of your article. Is it significant? Timely? New? Does it propose a different approach to an old problem? Are you sure that it is not just a restatement of an existing system or method? Is it really worth writing about? These are basic questions but they are important because they will help you determine, even before you start, whether or not you should take the time to plan and write the paper.

AUDIENCE

The next consideration is your audience. Who are you writing for? Are you trying to reach a specific, highly specialized group interested in higher mathematics and intricate technical details, or are you aiming your paper at the general public? What do you want your paper to do: educate, raise discussion, or simply impress your neighbors?

Honest answers to these questions are necessary. You have to establish the need for the paper, the type of persons it will interest, and the specific effect it should have upon them. Armed with this type of information you are then in a good position to start work on your paper.

THE SUMMARY

Now for the mechanics of writing. Set down in a few sentences exactly what you want to tell the reader.[1] Are you going to show him how to build a better mousetrap, or will you present a significant development in the state-of-the-art? State briefly the major points the article should convey. This is the first step to insure that your thoughts are going to move in a straight line.

If you have difficulties in summarizing what you want to tell the reader, you have a problem that must be resolved before going any further. Do not skip over this portion with the comment that you know perfectly well what you want to say and therefore you do not have to summarize it. Put it down now, even if it hurts to do so, and then examine what you are saying. It is revealing how often ideas that sound perfectly good when you think about them are not very clear or significant when expressed in black and white. Clarifying the thoughts at this point will make writing the complete paper easier for it will in essence define the boundaries within which the paper must be written.

THE OUTLINE

The next step in writing a technical paper is to make an outline. In your summary you have stated roughly what you want to tell the reader. Now you must decide how you plan to do this. What sequence of thoughts are you going to use? What points do you want to emphasize and what place in your paper are you going to do this? Generally an outline indicates the arguments or evidence you will use to prove to the reader that your new widget or idea is really good and that you know what you're talking about. This permits you to check quickly, before you start to write, whether the organization of your paper strengthens the major points you have written down in your summary.

There are, of course, many types of outlines. You can develop your paper chronologically, using time as the factor that determines your outline division. Or you may arrange your paper to evolve logically, starting from a premise followed by examples and/or arguments to prove the premise, finally

---

[1] Also refers to the audience at a symposium or convention.

leading to a conclusion that restates the premise. For instance, your outline may look like this:

1.0 Premise: System tests insure product quality.
2.0 Arguments:
   2.1 Lack of system tests permits shipment of defective systems.
      2.1.1 Example 1
      2.1.2 Example 2
      2.1.3 Example 3
   2.2 System tests reduce defects.
      2.2.1 Example 4
      2.2.2 Example 5
3.0 Conclusion: System tests insure product quality.

On the other hand, you may wish to approach your subject inductively, presenting examples from which you draw a conclusion. Or you can start with a concise statement of a problem or objective, show how the problem can be solved or the objective reached, and end up with a conclusion that shows the advantages of your procedure or the effects of reading your objective. Whatever method you use, the thing to remember is that the outline sets up the basic blocks and from these you can readily determine if your structure is sound: that is, that you are covering your subject fully, effectively, and without diverging from your main points.

One further consideration: you should estimate the length of your article or presentation. Technical articles seldom go over the 3,000 word limit. This is ten to twelve standard type pages typed double-spaced. Technical presentations longer than 30 minutes must be good to hold the attention of the audience. Using a speech rate of two minutes per page, that means a fifteen page text. Try then to estimate the number of pages each division of your outline will contain. A little work at this time will prevent you from widely overshooting your word limit on the first draft and save you some painful cutting later on.

REVIEW OF OUTLINE

Let us assume that you have made an outline containing four or five main parts. Now, keeping in mind what you stated in your summary, check these parts to see if they all support your original thought. Examine their sequence to make sure that each part is in the right place to produce this support. Sometimes one point will appear completely out of sequence. It is a good point, supplying important information, but left in its present position, it not only leads the reader away from what you want him to think, but it also confuses him. In that case you must relocate or eliminate the part in which it appears. Also check that you have enough parts to cover fully the points you want to make. Conversely, make sure that you are not swamping the reader with too many repetitive or unimportant facts. It is much better to present two or three important examples or to elaborate on a few representative ideas or principles than go into a mass of details. So choose your points carefully. Choose them for their significance and for their impact upon the reader.

ILLUSTRATIONS

Closely allied to an outline is a plan for illustrations. In a technical paper you will often need sketches, drawings, tables or photographs to illustrate your text. Unless you plan for these you may end up with your paper completed minus the necessary illustrations. Examine the main points in your outline and decide where an illustration will help clarify the thought. You may want a sketch of a test set-up, a drawing or a photographic slide of equipment. At another point you may want to present a chart to discuss data obtained in an experiment. A drawing may be required to show construction. You may even want to show how you derived a particular formula, although here a word of caution is in order. Unless your paper is aimed at a highly technical group vitally interested in your theoretical approach, leave out the complex formula. Remember, the size of a general audience decreases in direct proportion to the complexity of the mathematics. In any event, plan the illustrations for your audience. Make them large enough to be readable and, above all, make them add something to your text.

Once you have decided on the illustrations needed, you can make a separate illustration plan, or note the requirement beside the various points listed in your outline. The plan will act as a reminder that you must either do the work yourself or get someone else, like a presentation group, started so that you can have the illustrations ready with the paper. The notes in your outline remind you to refer to the illustrations as you write the paper. Some authors forget this. They add beautiful illustrations but they make absolutely no reference to them in their text. Thus, not only has the time spent producing illustrations been wasted but, worse, the help they could have provided to make the paper more interesting or convincing has been thrown away.

THE FIRST DRAFT

With your outline completed, you can start writing the first draft of your paper. At this time do not worry about language, grammar, or spelling. Just concentrate on developing the points you have indicated in your outline. Write down all the significant data you want to include under each point. If, as you write, you have some thoughts that do not specifically follow the outline but which you now think are pertinent, put them in. Essentially what you want to do is to fill in the blocks, adding the details, the examples that are needed to make your thoughts clear. This phase can often be done rapidly, writing at top speed until the complete draft of the article is finished. Of course some people may find that they work more slowly, completing only parts of the paper at one sitting. It does not matter. The goal is to get the whole draft written as quickly as possible, without deviating too much from the outline, so that for the first time you can see the complete article.

Here the question may arise, how long should my draft be? It should be long enough to cover all the points listed in your outline and, if possible, it should be close to your estimated length.

The draft is generally written or typed double or even triple-spaced to leave plenty of room for changes and corrections. And while it is not imperative that it be typed, the cleanly typed copy will make it easier to read. It also gives you a better feel for the impact the full article will have upon the reader.

REVIEW THE DRAFT

When the first draft has been completed it is worthwhile to read it as an entity. Reading it aloud helps or, if you have a tape recorder available, tape it and then listen to it. At this time all you should do is correct the most obvious errors in English and note where you need to add or delete information. After that lay the draft aside for a week or more. This is done to gain perspective, for as long as you are close to the writing you cannot truly evaluate it. The time delay helps make you forget the immediate details so that when you read the paper once more, you can edit it more objectively.

EDITING

Editing is one of the most important phases of writing a paper. Don't be scared by the word "editing." It means nothing more than examining your paper slowly, critically, and making the necessary changes to smooth the language, insure that your thoughts are expressed clearly, and that you make your points well. This is the same type of review accorded the outline except that now you are checking the details of your writing to see how closely they follow your plan. Granted that this is hard work, but if you want to be effective and produce a good paper, it must be done carefully.

In editing, you assume that you are the reader and that you must be interested, entertained, taught, or convinced by what you read. Your paper should accomplish this, first, by making it easy for the reader to read and understand your thoughts and, second, by the weight of the logic and evidence presented to support your conclusions.

To check on the first point, read the article again for flow of language. Do the sentences flow smoothly into each other? Do the words sound right? Are they simple and clear? Sometimes the thought can be confused by too long a sentence. Chop the thought into easily digested doses by using short sentences, varied occasionally by a longer one where detailed explanations make more words necessary. If desired, tape the article again and listen to the playback. The awkward constructions and muddled sentences will stick out and can readily be corrected.

As you work to smooth the language also examine the text from the point of view of reader interest and understanding. Is your first sentence startling? Does it arouse curiosity? Will it make the reader want to continue reading? Then look at the way you express your thoughts. Are you really saying what you want to say in the shortest and most powerful way possible? This brings up one of the toughest editing tasks. If any portion of your article appears to drag, if there is any extraneous material, or if any part is out of proportion to its weight in supporting a particular point, don't hesitate to cut. I realize that the article is your creation, and that every part of it is sacred, but you have to be ruthless. A pet word, showing off your knowledge of technical terms, but which is vague or confusing must be yanked out and replaced by a simpler, better-known term. A good rule is to cut your draft by 20 percent. You say that is impossible, but you would be surprised how much it will improve your article.

Also, in presenting your thoughts to the reader, do not ever assume that he knows as much about the subject as you do. Too often trade names, abbreviations, and functions are mentioned without explanation on the assumption that the reader is in the same business and therefore is totally familiar with the terms. At other times it is strictly an author oversight. Being completely at home with the subject, the author overlooks the fact that another person lacks the detailed knowledge, or even some of the fundamentals, which he so casually omits. Remember that it is better to err slightly on the side of details, particularly for complex technical subjects, than to leave the reader with a lot of gaps which he has neither the time nor the inclination to fill.

When you are reasonably sure that the reader will understand what you are saying, then you must examine how well your article can convince him. If you have done your homework well and considered this phase in your outline, the basic logic should be there . Now all you need to do is check that your development of each point is effective. For instance, if you want the reader to agree that system tests insure product quality, every statement you make, every example you give, must support this contention. The evidence you present may be positive, showing the advantages of running system tests; it may indicate the disadvantages of not performing them; it may indicate the effects of engineering tests, of field tests. Your argument may proceed from the known to the unknown; it may evolve logically or chronologically, but to be pertinent, every sentence written must proceed to stress, to restate, perhaps by examining different aspects of the subject, to develop and to extend the basic thoughts you indicated in your summary until, by the weight of evidence and logic, the reader is convinced and agrees with your conclusions.

In this task make sure that your examples are good, for a vivid example not only can clarify a statement and evoke reader interest but often can act as the clinching argument. For instance, if I say that lack of a system test causes failures, the statement by itself does not have much weight, but if I add: "for example, when flight testing missile Y in the desert, we omitted the system test to meet a scheduled commitment. But we forgot one thing, the one thing that nearly killed us . . . etc.," then I gain reader interest and can more easily make my point.

One important consideration: if you have any doubt about any portion of your article, even when it is finished and ready to be sent out, do not hesitate to rewrite. Often the writer will let the article go even though he feels that the language is fuzzy or the thought is not completely clear. With the normal reluctance to do extra work he will rationalize that the reader will understand his point. Not so! If you have any qualm about any part of your article you can be sure that the reader will stumble over it. So rewrite the portion to make it better. You may have to struggle to do so, but once it has been done you will feel that it was worth the effort.

FINAL DRAFT

Now you should be ready to write your final draft. This is your calling card to the editor so you have to make it good. The draft should be typed double-spaced with a good ribbon on 8½ by 11 inch white bond. Colored paper might look pretty but it does not impress the editors.

Start half-way down the first page with your title, name, and under it, your present affiliation. For example:

The Hybrid Life of the Electron
by
John C. Shufholder
Senior Engineer
XTA Corp.

The title should be fairly short and so conceived as to arouse interest or pose a question. It must cause the reader to pick your article from the others offered him.

In typing the text, leave a margin on each side, wider on the left, so that the editor has plenty of room to write his comments or instructions to a printer or typist. Also, do not forget to make a carbon copy. You might need it in case the original is lost. Now as you type you may want to make minor corrections. Go ahead. The aim is to make this version as good as possible. But be sure that the copy is clean and without typing errors. Remember that a sloppy, over-corrected, and dirty manuscript creates the impression that the author thinks and works the same way. This may not be true and your paper may be the work of genius, but do not handicap yourself by adversely conditioning your first reader.

At this time you should examine your final illustrations with the following points in mind:

1. Are the illustrations adequately tied in to the text?

2. Do they really add understanding or weight to the paper?

3. Are they too complex? Or is the printing or drawing too small to be effective?

Do not let someone else handle this phase on the excuse that he knows more about illustrations than you do. This is your paper and, as the author, the quality of everything that goes into it is your responsibility.

Before submitting your final draft it may be worthwhile to ask a qualified person to review and criticize it. At this time ignore congratulations on your writing ability or praises of your technical virtuosity. What you are looking for are answers to the following:

1. Am I defining my problem or my purpose well?

2. Is my development clear and understandable?

3. Do I make my point(s)?

If the answers to these questions are in the affirmative and you have generally complied with the previous suggestions, you should be on your way toward having your paper published or accepted for presentation.

DISCUSSION AND ACTIVITY

1. After considering the need for writing a paper and the intended audience, Farkas describes the steps in writing the paper. What are the steps? What is achieved in each step?

2. Most advice about editing (or revising) and proofreading explains what the writer should do but is short on specific advice on how to actually do it. Experienced writers, like veteran linebackers, develop a kind of peripheral vision that enables them to look for a lot of things simultaneously. Most beginning writers cannot check on several

things at the same time. What approaches to revision and proofreading might beginning writers take to insure that their work is revised and proofread adequately?

3. Farkas's first suggestion is to establish the need for any paper you are planning to do. That is, you should determine why the information in your paper is important for your reader to know. Select some topic (such as the differences between a rip saw and a cross-cut saw, or how to make homemade granola, and so forth) and write a paragraph explaining why you think the reader should possess the information.

Selection Sixteen

Selecting and Writing to the Proper Level

JOSEPH RACKER

*When he wrote the following article, Joseph Racker was vice president
and chief engineer of the Caldwell-Clements Manuals Corporation,
New York. Racker explains how to determine the proper level at which
material should be presented.*

Generally in any form of objective writing, and particularly in technical
writing, the writer tries to convey information in the most direct and concise
manner. To do this, the writer assumes that the reader has a certain amount
of basic information and that this information does not have to be repeated.
As a result the writer may omit a great deal of pertinent data, relying upon
the reader to supply this data. This point can best be understood by the two
simple examples shown in accompanying Tables 16-1 and 16-2. Summarizing: the *writing level* actually represents the level, or general background, that
the *reader* must have to understand the writer.

IMPORTANCE OF SELECTING THE PROPER LEVEL

Technical writing is a form of instruction and many of the techniques that
represent good teaching practice also apply to good writing practice. It must
be remembered that readers are human and subject to human limitations. Do

Table 16-1   Measurement of a d-c voltage

Consider the instructions: "Check that the voltage at pin 5, V101 is +160 volts, ±10%. If it is not, adjust R101 until this reading is obtained."

In order to perform the instructions indicated in this statement, the reader must know and fill in the following bits of information:

1. The locations of pin 5, V101 and R101.

2. That voltage must be measured between pin 5, V101 and ground.

3. The location of ground.

4. That a d-c voltmeter must be used and the type that should be used.

5. The procedure for using the d-c meter.

6. That a reading of between +144 and +176 volts, d-c, is acceptable.

7. That R101 must be adjusted if the reading is less than 144 or more than 176.

8. How to adjust R101.

9. That if it is necessary, R101 should be adjusted for a reading of +160 volts, dc.

Readers that can fill in all of this information can perform the instructions given without any difficulty. Readers that do not have all of this information may find it difficult, if not impossible, to perform this instruction correctly.

not assume that if the reader is sufficiently interested in your paper he will "puzzle out" all of the information you have failed to supply, or conversely that he will read through a great deal of extraneous material to find an occasional bit of useful information.

Generally a reader will not read material that is above his level. Obviously if he is required to supply information that he does not have, he cannot follow the text. Few readers will read several other background books or articles in order to understand the one you have written. Most readers will dismiss the text that is above their level as being incomprehensible.

Most readers will absorb material best that is written slightly below their level. However, if it is considerably below the reader's level, he will tend to skim over large portions of it and generally will miss some vital information.

The over-all result is that he will often end up with the wrong results or misleading information and attribute it to the poor technical background of the writer. Human nature is such that a reader will often assume that material written to a much lower level is also inaccurate or inadequate.

The ideal level is the one that allows the reader to supply all information

Table 16-2   Logarithmic amplifier response curve

Consider the instruction "Check that the amplifier response is linearly logarithmic (within 1 db) over an 80 db range."

In order to perform the instructions indicated in this statement, the reader must know and fill in the following bits of information:

1. What the expression "linearly logarithmic (within 1 db) of an 80 db range" means.

2. What test equipment is required to measure such a response.

3. How to connect and operate such test equipment.

4. The starting point of the 80 db range.

Readers that can fill in this information can perform the instructions given without any difficulty. Readers that do not have all of this information will find it *impossible* to perform this instruction correctly.

---

readily available to him and provides only information that is new or not fresh in his mind. Facts that are generally available but not used frequently (and likely to be forgotten) should be included in this ideal level of writing.

As in every ideal situation, it may be approached but can never be reached. The writer could not know all of the information that is readily available to the reader. Even if he had this information, many readers must be considered and each has his own level. The best that the writer could hope to do is to write at a level that is satisfactory to the average reader.

As an aid in writing to the average reader, five levels of technical writing, and the assumed background of the average reader for each level, will be defined. Then I will describe some general rules that should be followed for each level. Before defining these levels, I would like to emphasize a general rule for writing at any level.

GENERAL CONSIDERATIONS IN WRITING AT ANY LEVEL

The important consideration in writing at any level is to be fully cognizant of all the information necessary to understand or carry out the written material. When the material involves instructions, this can readily be accomplished by actually performing the instructions and carefully noting every step that is taken. If this cannot be done because the equipment is not available, then the writer must rely upon his experience and visualize every step that is taken.

For example, the statement "Tune Coil L8 for maximum output" may be as simple as it sounds or may be a very critical adjustment requiring special instructions. The writer should know all of the steps necessary to tune L8 properly.

It is equally important, though not as easy to visualize, to determine all the information necessary to follow papers on equipment design and theory. A common occurrence in this category is when an author gives one equation and then states that another equation can be derived from this first one. In many cases the derivation is arrived at by complex mathematical operations and possibly by elimination of some terms deemed negligible for the problem being considered. Certainly the author should be fully cognizant of all the steps and assumptions necessary to derive one equation from another.

I am not trying to state at this time that all information must be used. At some levels of writing omissions of information are justified. However, before an author can determine what information can be omitted he should be *fully cognizant of all the information necessary to understand the subject matter.* This is the point being emphasized at this time. The selection of what information is or is not required for a particular level is often far easier than determining all the information required.

Now we come to the definitions of five levels of technical writing. First I will define them, then I will describe how to write to each level. The five levels are as follows (See Table 16-3).

DEFINITION OF OPERATOR'S OR NON-TECHNICAL LEVEL

In an operator's or non-technical level of writing, the reader is assumed to have no specialized training in the subject matter of the text. Note that this does not necessarily mean that the reader does not have any technical background at all. Examples of writing that fall into this category are as follows:

1. Describing the operation or maintenance of an equipment to personnel who have no training on this or similar equipment (for example, a military trainee with no technical training).

2. Preparation of a handbook for instructing pilots on the operation of an electronic auto-pilot. While the pilot has an extensive technical background, he may have little or no knowledge of the theory and operation of electronic automatic landing systems. Consequently, from the viewpoint of his knowledge of electronic equipment, the pilot is considered a non-technical person.

Table 16-3   Five levels of technical writing

1. Operator's or Non-Technician level.
2. Field or Technician's level.
3. Depot, advanced Technician's or Junior Engineer's level.
4. Engineer's level.
5. Advanced Engineer's or Scientific level.

3. Preparing a brochure or proposal directed at high-level corporation or military personnel who are either not interested in or not qualified to understand the technical details of a project. However, they are interested in the important over-all results such as reduced costs, weight, or other advantages that they can readily evaluate and in a simplified explanation of how these advantages are obtained.

DEFINITION OF FIELD OR TECHNICIAN'S LEVEL

In a field or technician's level of writing, the reader is assumed to have special training on a specific equipment or class of equipment with a specified number of tools or test equipment. Generally the reader has been trained to maintain the equipment while it is in the field, home or airport. Examples of personnel that fall into this category are:

1. A television repairman who services calls in the home. Generally he is capable of repairing simple, common troubles, making some adjustments, and removing major assemblies.
2. An aircraft mechanic who checks the operation of aircraft on a routine maintenance basis at the airport and makes periodic adjustments and minor repairs.
3. A technician who services electronic or mechanical equipment in the field, by following routine preventive and corrective maintenance instructions such as lubrication, adjustments, and replacement of readily accessible and removable parts.

DEFINITION OF DEPOT, ADVANCED
TECHNICIAN'S OR JUNIOR ENGINEER'S LEVEL

In a depot, advanced technician's or junior engineer's level the reader is assumed to have extensive background information, either in experience or schooling, on the subject matter. Generally the reader is capable of thoroughly understanding the theory of operation of the equipment and can select the proper tools or test equipment necessary to maintain the equipment when it is sent to the depot, shop or factory or aircraft hangars. Examples of personnel that fall into this category are:

1. The factory or shop television serviceman who repairs and aligns sets that cannot be handled by the field man.

2. Technicians who overhaul electronic or aircraft equipment.

3. Technicians and junior engineers who build, test, and "debug" developmental models of equipment. These personnel generally can design test setups and make minor design modifications.

DEFINITION OF ENGINEER'S LEVEL

In the engineer's level, the reader is assumed to have an engineering degree, or equivalent, and at least several years experience as a practicing engineer. Generally at this level the primary emphasis is in the development of new equipment or techniques rather than in the maintenance of existing equipment.

DEFINITION OF ADVANCED ENGINEER'S OR SCIENTIFIC LEVEL

In the advanced engineer's or scientific level, the reader is assumed to have an advanced engineering degree or extensive experience in a particular field. Generally at this level, the primary emphasis is in theoretical calculations, new concepts, or basic research.

WRITING TO OPERATOR'S OR NON-TECHNICAL LEVEL

I will describe some general rules to be observed in writing to an operator's or non-technical level. (See Table 16-4.)

1. *Include all information.* Do not expect the reader to supply any technical information in order to understand the text. Thus, in the example given in Table 16-1 covering taking of a voltage reading, all 9 items of information would be supplied.

2. *Back text with simple, pictorial illustrations.* Extensive use of illustrations to emphasize and clarify text is highly recommended. Illustrations should be simple and duplicate the physical appearance of the equipment. Schematic or symbolic representations should not be used.

Table 16-4   General rules for writing to operator's or non-technical level

1. Include all information.
2. Back text with simple, pictorial illustrations.
3. Use step-by-step procedures.
4. Keep reference to other procedures or books to a minimum.

3. *Use step-by-step procedures.* Whenever possible, instructions should be presented in short, distinct steps. When a large number of instructions are involved, the operational steps should be subdivided into procedural steps. For example:
   A. Turning Meter On
      1. Place meter power cord into a-c outlet.
      2. Place meter POWER switch in the ON position.
      3. Allow meter to warm up for 30 seconds.
   B. Setting of Meter Controls Prior to Operation
      1. Place AC-DC switch in AC position.
      2. Place RANGE SELECT switch in 100 VOLTS position.
      3. Adjust ZERO knob for a zero reading on meter.
   C. Connection of Meter to Set
      1. Locate terminal marked AUDIO OUTPUT at back of set.
      2. Connect red lead of meter to AUDIO OUTPUT terminal.
      3. Connect black lead to chassis.

4. *Reference to other procedures or books should be kept to a minimum.* As much as practical, all steps necessary to perform an operation should be listed in text. References to other procedures or manuals such as "turn equipment on as described in manual covering meter" or "repeat steps a to c, paragraph 1-15" or "turn power off" should be used sparingly. Unless it becomes unduly repetitious, include all steps given in the manual covering the meter, literally repeat steps a to c, and provide the procedure for turning power off.

WRITING TO A FIELD OR TECHNICIAN'S LEVEL

I will now describe some general rules to be observed in writing to a field or technician's level. (See Table 16-5.)

1. *Specify test equipment normally available in the field, when possible; provide detailed procedures on special test equipment.* A widely used method of determining where field level maintenance ends and depot level maintenance begins is to consult the list of test equipment available in the field. In military applications a list of field test equipment is usually made up for each equipment; in

Table 16-5   General rules for writing to a field or technician's level

1. Specify test equipment normally available in the field, when possible; provide detailed procedures on special test equipment.

2. In theory of operation, provide sufficient information to permit proper performance of maintenance but do not repeat basic theory or include detailed theory not required for maintenance.

3. Use graphs or nomographs instead of equations when possible.

4. Repetition of instructions is not necessary, reference can be made to other paragraphs or manuals.

---

commercial applications this list is determined by common practice. If maintenance can be performed with test equipment available in the field, it is considered field maintenance. All maintenance requiring test equipment not available in the field is generally considered depot maintenance. It is not necessary to provide detailed operating and test setup instructions on field test equipment. However, for the exceptional case where it is necessary to use special test equipment in the field, detailed information should be supplied on how to set up and operate this equipment.

2. *In theory of operation, provide sufficient information to permit proper performance of maintenance but do not repeat basic theory or include detailed theory not required for maintenance.* In preparing maintenance information, such as trouble shooting charts, it is impossible to list every possible trouble that may occur in the field. Only general procedures can be given together with some common causes of failure. To enable the technician to deal with many of the specific problems he will face, a theory of operation section is generally included. This section should not repeat basic theory, or describe details of theory that will not aid in the maintenance of the equipment at the field level. For example, in radio equipment, it would not be necessary to describe the operation of a conventional amplifier or rectifier, nor would it be necessary to describe the detailed operation of a sealed motor which, if inoperative, would have to be replaced entirely. The theory section should include a description of any special circuits and indicate the key components of all circuits. For example, stage V101 is an amplifier which utilizes a feedback resistor, R101, to reduce distortion and improve the stability of the amplifier. Or, V102 is a multivibrator whose repetition rate depends primarily on the r-c network of R103, C101, and R102. The effect of any variable components, which the technician must adjust, should be particularly emphasized. For example, the audio output is determined by the setting of potentiometer R105 (AUDIO OUTPUT) which varies the amplitude of the signal applied to the grid of

audio amplifier V103. Liberal use of simplified schematics should also be employed.

3. *Use graphs or nomographs instead of equations when possible.* In some instances, it is necessary to use mathematics to explain theory or perform maintenance. Generally the field technician can be expected to understand some simple mathematics, such as sine and cosine functions, but even simple mathematical formulas and certainly those involving higher mathematics should be supplemented with illustrations that enable solution of the equations by means of a straight edge. . . .

4. *Repetition of instructions is not necessary, reference can be made to other paragraphs or manuals.* At this level it is not necessary to repeat information included in other procedures in the book or those contained in other manuals normally available in the field.

WRITING TO A DEPOT, ADVANCED TECHNICIAN'S OR JUNIOR ENGINEER'S LEVEL

I will now describe some of the general rules that should be observed in writing to a depot, advanced technician's or junior engineer's level. (See Table 16-6.)

1. *The operation of any available military (for military applications) or commercial test equipment need not be covered.* At this level it can be assumed that any available piece of test equipment can be procured and that the reader can become, or is, familiar with the operation of this equipment. Furthermore, it may be assumed that the reader can select equivalent test equipment that he may have available (modifying it when necessary to make it equivalent to the equipment specified).

2. *Special or fabricated test equipment can be specified and should be described.* Any special test equipment that is available can be specified and if such equipment is not available, the fabrication of such equipment should be covered. Any operating instructions that are unique to such special test equipment should be described.

3. *Describe all special disassembly and assembly techniques, supplement such procedures with diagrams.* Complete disassembly and reassembly procedures are often included in books written at this level. While it is not necessary to describe obvious disassembly or reassembly steps, any special techniques should be covered. That is, if it is just a matter of loosening some nuts or unscrewing some screws to disassemble equipment, a detailed procedure is not required. However, if it is necessary to set a gear in a specific position before a holding screw is loosened in order to slip the gear out, this instruction should be described. Diagrams, preferably exploded views, which illustrate the disassembled equipment, identify each part, and indicate by index number the

Table 16-6   General rules for writing to a depot, advanced technician's or junior engineer's level

---

1. The operation of any available military (for military applications) or commercial test equipment need not be covered.

2. Special or fabricated test equipment can be specified and should be described.

3. Describe all special disassembly and assembly techniques, supplement such procedures with diagrams.

4. The complete theory of operation of the equipment must be included; conventional circuitry may be covered by a block diagram description.

5. Use graphs and nomographs instead of equations when possible.

6. Repetition of instructions is not necessary, reference can be made to other paragraphs or manuals.

---

order of disassembly, should be used to supplement the text. Such a diagram should be provided for even simple disassembly procedures because after an equipment is disassembled the reassembly may not be apparent, at least to the extent of replacing all the lockwashers, using the proper sized nuts and bolts in the right places, etc.

4. *The complete theory of operation of the equipment must be included; conventional circuitry may be covered by a block diagram description.* The theory of operation should cover every circuit and include a complete set of diagrams to permit servicing every item in the equipment. Text description of all conventional circuits can be on a block diagram basis such as "the receiver consists of an r-f stage V101, mixer-oscillator V102, three i-f stages V103 to V105, detector V106, and audio amplifier V107." However, a detailed schematic of all of these circuits should be given as well as a description of any circuits that are unique to this equipment.

5. *Use graphs and nomographs instead of equations when possible.* The comments made on this subject for field level also apply at this level.

6. *Repetition of instructions is not necessary, reference can be made to other paragraphs or manuals.* The comments made on this subject for field level also apply at this level.

WRITING TO AN ENGINEER'S LEVEL

I will now describe some of the general rules to be observed in writing to an engineer's level. (See Table 16-7.)

Table 16-7   General rules for writing to an engineer's level

1. Start text with basic information that reader must know.

2. All information should be objective.

3. Information should be complete.

4. Simple calculus can be used freely; advanced calculus should be avoided, if possible, or referenced in an appendix if it must be used.

5. Define all terms in equations; use standard symbols when possible.

6. Use same units throughout text, when possible, or clearly indicate when new units are being used.

---

1. *Start text with basic information that reader must know.* A brief review of the basic information that the reader should know should be included in one of the early paragraphs. When limitation of space does not permit repetition of basic information, then an indication of what basic information is required and where it may be obtained should be included.

2. *All information should be objective.* Material written at this level should be objective and should not include digressive and unnecessary information. For example, a description whose purpose it is to gain acceptance of a new equipment by potential users should emphasize performance characteristics and need not detail design problems unless they are important to the evaluation of the equipment performance.

3. *Information should be complete.* All information necessary for the understanding of the material should be included in the text either directly or by reference to other source material. The terms "it can be shown that" should be used only in conjunction with a reference that does show it. All assumptions, approximations, or short cuts used to arrive at conclusions should be stated. In other words, it should be possible for any engineer reading this material with all referenced source material to fully understand and evaluate the material presented. Many an author has been embarrassed by the question, "How did you arrive at this conclusion?" coming from a reader several years after a paper was written. The conventional reply is "I knew it when I wrote the article, but now I must look at my notes to refresh my memory." If it is necessary to consult the author's notes in order to understand the text material, then the material has not been properly written.

4. *Simple calculus can be used freely; advanced calculus should be avoided, if possible, or referenced or placed in an appendix if it must be used.* Equations involving mathematics no higher than simple integration or differentiation can be used freely. However, if advanced differential equations or integrations are involved, it is advisable merely to state the over-all conclusion and refer to

another text for the derivation of this conclusion (when it is derived in another text) or include the derivation as part of an appendix. Of course, such mathematics should only be included when it is necessary to achieve the objective of the article. Some authors believe that it adds to their prestige to include complex mathematical formulas in their papers. This is *not* true if these formulas are not necessary to accomplish the objective of the article. In this latter case, it merely provides a major obstacle to the understanding of the material and often completely discourages readers from reading the paper.

5. *Define all terms in equations; use standard symbols when possible.* When using equations clearly define each term in the equation including its units: for example, v is the velocity in feet per second. When possible use the standard symbol, as defined by the IRE or equivalent engineering authority, for each term.

6. *Use same units throughout text, when possible, or clearly indicate when new units are being used.* A great deal of confusion can result when the author goes from one set of units to another set; for example, inches to centimeters, grams to ounces, degrees to mils. If possible all of the information should be presented in one set of units. If this is not possible, then the fact that a conversion is being made and the conversion factor used (if it is not well known) should be clearly specified.

WRITING TO AN ADVANCED ENGINEER'S OR SCIENTIFIC LEVEL

1. *Introduction should reference sources where basic information can be obtained.* At this level of writing the author generally cannot start with basic information and keep text within reasonable limits. He should reference source material where this basic information can be obtained.

2. *Higher mathematics can be used freely; however, it should be placed in footnotes or the appendix when possible.* Any mathematics necessary to develop the subject matter of the paper can be used. When possible, detailed derivation of an equation involving higher mathematics can be placed in a footnote or appendix.

3. *Information should be complete.* The comments made on this subject for an engineering level also apply at this level.

4. *Define all terms in equations; use standard symbols when possible.* The comments made on this subject for an engineering level also apply at this level. However, since a great number of equations may be used and some terms repeated several times, it is convenient to define all terms at the beginning of the article.

5. *Use same units throughout text, when possible, or clearly indicate when new units are being used.* The comments made on this subject for an engineering level also apply at this level.

DISCUSSION AND ACTIVITY

1. How do most readers react when they try to read something that is above their level of comprehension?
2. At what level do most readers best absorb material?
3. Racker identifies five levels of technical writing, explains the assumed background of the average reader for each level, and discusses some general rules for writing to each level. Choose some topic and describe the different audiences you might write to and how your writing would differ for each audience.
4. In the following selection by H. R. Clauser and in the suggested further readings at the end of this section by Gordon Mills and John Walter and Thomas E. Pearsall are discussions of audience analysis. After studying Racker, Clauser, Mills and Walter, and Pearsall, write a brief report explaining the similarities and differences among the four pieces in terms of audience levels, characteristics of each audience level, and adaptations that should be made in writing to each audience level.

Selection Seventeen

Writing Is for Readers

H. R. CLAUSER

H. R. Clauser was editor of Materials in Design Engineering *(now* Materials Engineering*) when he wrote the following article. Clauser recommends that writers keep the rules of good reading in mind when they write.*

Reading is an indispensable part of written communication. To communicate, our words, phrases, and sentences must be read and understood. Writing that does not meet this requirement is noise.

It follows, then, that it is the reader's reception of our writing that measures the extent to which we successfully communicate our ideas; thus the reader is a proper and important person to study when we set out to improve our writing. So it is he, the reader, rather than you, the writer, who will be the principal subject of this article.

In recent years reading courses have become almost as popular as courses on how to win friends. And rightly so, for they can improve reading speed and comprehension. But in one way, reading courses are based on a partially false assumption—that is, that all writing is good writing and that writers are aware of the rules of good reading. We all know that this is far from being the case.

So, to help our readers and make it possible for them to practice what they have learned from reading courses, we should write in accordance with the rules of good reading. If we follow this practice, we will quickly discover that the rules also help us, the writers, for as it turns out, the rules of good reading are identical with those of good writing. Let's study the rules for effective reading.

STWP Review 8 (January 1961), 12-17. Permission granted by The Society for Technical Communication, 1010 Vermont Avenue, N.W., Washington, D.C. 20005.

READING AND WRITING WITH A PURPOSE

The following statement introduces one of the chapters in a course on reading: "When an author writes an article or a chapter, he has a purpose in mind for conveying his thoughts to others. . . . When you are about to read a selection, you should decide what that purpose is, and adjust your reading to it." This is sound advice, and I'm sure many readers, whether or not they have taken a reading course, try to apply it.

But, unfortunately, readers are frequently double-crossed by writers whose writings seem to have no discernible purpose. Reading such writing is like retrieving a ball of yarn that is aimlessly rolling and unwinding on a patch of uneven ground. The reader does not know what to expect. He cannot determine the level or pattern of the thoughts or information the writer wants to convey. Consequently the reader may become discouraged and give up; or, having retrieved the unraveled ball of yarn, he may wonder why he did.

So, for effective communication there must be a definite purpose behind our writing to give it form and a sense of direction. To determine what this purpose is, we must ask these two questions before beginning to collect or select our material: What is it for? and Whom is it for?

Answering these questions will go a long way toward giving your writing a sense of unity and focus. There are often several answers to the questions. You may have several aims in mind. The solution, then, is not to ignore the questions and try to serve all possible purposes. Difficult as it may be, you must still arrive at answers that will define the purpose of your work for the reader.

There is another important reason for answering the foregoing questions. Writing with a definite purpose not only benefits the reader, but also helps you, the writer. As you will see later, it serves as the benchmark, the basic guide for your writing task. It sets the tone, the pace, and the technical level; it helps determine the amount of detail to be included; and it suggests how the writing should be composed.

READING AND WRITING IDEAS

One of the first things the students in reading courses learn is to read ideas and not just words. They are taught to master the art of searching for and quickly grasping the author's ideas.

The key to helping the reader in his search for our ideas is the paragraph. The principles of good reading teach him that each paragraph viewed as a whole represents one and only one unit of thought or idea. So the writer should approach each paragraph as if it were the only writing before him at that moment, and it should be written with these two questions in mind:

1. What is the basic *subject* or thing I want to cover in this paragraph?
2. What is the important *idea* or thing about the subject I want to convey?

Paragraphs constructed using these questions as a guide will make it easy for the reader to spot the basic thought and to separate it from supporting, but often subordinate, detail.

In every piece of expository writing, there is supposed to be a hierarchy of ideas and subjects, and the student taking a reading course is taught to uncover that hierarchy and quickly pick out the main ideas. Thus, to help him in this task, our writing should clearly distinguish between the importance of various ideas we are trying to convey to him. There are a number of ways to do this. A well chosen title for an article, a report, or a memo can often give the reader at least a good preview of the principal idea or ideas covered. More commonly, the subheads can function as statements of either the main ideas or subjects and, at the same time, show the structure of the written piece. For example, in this article the subheads represent the major subjects covered. Other familiar methods are abstracts, summaries, and lists of major points covered. The accompanying box, for example, highlights the main ideas of this article.

HERE ARE THE HIGHLIGHTS

• In written communication, no one plays a more important part than the reader.

• The rules of good reading are identical with those of good writing. In terms of writing they are:

1. Write with a purpose.
2. Write ideas, not just words.
3. Write paragraphs as units of thought.
4. Highlight the main ideas.
5. Include only relevant detail.
6. Select amount and kind of detail to suit the audience.
7. Relate detail to the main ideas.
8. Permit conscious inaccuracy when it leads to better understanding.

A method not so common but which could be used more often to advantage is leading off each major section of the paper with a statement of the main idea. The idea can be emphasized by underlining it. In printed matter the main ideas can be set in boldface type, as is shown in Figure 17-1.

READING AND WRITING DETAILS

Perhaps the most difficult part of reading and, likewise, of writing, is dealing with detail. Often it is relatively easy to set down the main ideas of a piece of writing. But when that has been done, writer's (or reader's) cramp frequently sets in. We either feel that there is nothing more to say, or we feel hopeless and indecisive in facing the mountain of detailed material before us.

In dealing with detail, our first concern is to avoid burdening the reader with unnecessary detail. Out of the mass of facts on hand, we must sift out that which is relevant. "What has this to do with it?" This is the question we must constantly ask because irrelevant details seldom do any good and may do some harm.

Indeed, one of the unkindest tricks the writer can play on the reader, particularly one who has taken a rapid reading course, is to suddenly present him with an isolated, irrelevant fact. In the paragraph below, the last sentence is such an isolated fact, with no relation to the rest of the paragraph. Examination of the entire article showed that it also had no connection with anything else in the article. This example illustrates a particularly dangerous type of irrelevancy. Because "hardenability," the irrelevancy, in some instances could be related to the subject of "carbide content," the reader is doubly confused.

A further control of carbide content can be obtained by composition variations. The carbide content is increased by the addition of chromium or by lowering the carbon equivalent, and it is decreased by the addition of copper or nickel. *Hardenability is significantly increased by the addition of molybdenum or chromium and is moderately increased by the addition of copper or nickel.*

Sprinklings of such irrelevant facts through technical papers and reports is common in technical writing. Perhaps the most common cause is that most of us do some thinking as we write and sometimes even write in order to think. Therefore it is not unusual that, from the many associations generated by our thinking, some unwanted ones will drop unnoticed into our paragraphs. The cure is to seek them out when you revise and never to succumb to the temptation to write the way you think or talk.

Here is a

comprehensive survey of

Permanent
Magnet
Materials

- *Magnet steels* • *Alnicos*
- *Other magnetic alloys*
- *Ceramics* • *Fine particle magnets*

by **Robert J. Fabian**, *Associate Editor, Materials in Design Engineering*

■ Permanent magnet materials can be used in a wide range of applications where a self-contained source of potential energy is needed. Their use ranges from simple door latches having a high holding force, on up to complex motor stators having a strong, inherent magnetic field. A great number of permanent magnet materials have been developed to meet the needs of such varied applications. The purpose of this article is to describe the various metal and ceramic materials that are available today.

In the past, the magnet steels described first below were the only permanent magnet materials available to the designer. Although they are still used to a limited extent today, they have been replaced in many applications by the newer permanent magnet alloys and ceramics which possess superior magnetic properties.

Magnet steels

are generally low in cost and are available in a wide variety of shapes and sizes. Although suitable for some applications, they are not used where high magnetic strength, small size or light weight is required.

At one time the magnet steels were the only permanent magnet materials available to the designer. Although they are still used to a limited extent today, they have been replaced in many applications by the newer permanent magnet alloys and ceramics which possess superior magnetic properties.

Figure 17-1. Highlighting the main ideas

HOW MUCH, WHAT KIND OF DETAIL?

Besides having to decide what is relevant to the subject about which we are writing, there are other problems to face. One of these is deciding the amount and kind of detail needed.

The amount and kind of detail needed to amplify our main ideas is closely related to the audience. Every field, every subject, consists of several stages of sophistication. And the stage that your audience occupies will largely determine the kind and amount of detail you can safely include. For example, in the general field of science, there are five stages of sophistication (see Figure 17-2), according to Bachelard in his book, *La Philosophie du Non*. The double lines separating the stages represent the real barriers that exist be-

STAGE 1	STAGE 2	STAGE 3	STAGE 4	STAGE 5
Primitive Realism	Empiricism	Classical Science	Modern Science	Advancing Science

Figure 17-2.   Stages of scientific sophistication

tween stages, and when writing in the general science field, we must respect them.

Although it is relatively easy for a person in one stage to move to a lower stage and back again, it requires a good deal of effort to hurdle the barrier into the next higher stage. Thus, if the audience is largely in Stage 3, the detail you present must be largely of the Stage 3 variety or less. Or if you are addressing a Stage 4 subject to a Stage 3 audience, only a small amount of detail can be successfully communicated. And, according to W. M. Thistle,[1] to try to transmit any detail at all back through two or more of the stages is almost an impossible task.

What should be done when different amounts of detail are desired by various members of a mixed audience? Should the paper, report, or article contain practically everything—the laboratory procedure, the test set-up, the reasoning steps, the rejected alternatives, and answers to all questions or criticism that might be raised? If we include all such details in the conventional manner, we reduce the impact of our major findings and run the risk of losing important readers. On the other hand, if we omit some of the detail, we can be criticized for being incomplete or for not fully substantiating our findings.

It is a difficult decision to make. But rather than playing it safe and always including "everything," the problem should be faced each time we have a writing job to do. In many cases the problem can be solved by making use of appendixes. Another way is to treat some of the less important detail as optional reading. This is easily done in printed material by enclosing the optional reading in a box, as shown in Figure 17-3. When other methods of reproduction are used, the technique of optional reading is more difficult to apply. Nevertheless, it is still often feasible to insert boxes even in typewritten pages.

HOW ACCURATE THE DETAIL?

Closely related to the amount and kind of detail is the question of how accurate detail must be when conveying ideas in one stage to readers who are in a lower stage of sophistication. Many scientists and technical men insist on

manganese steels, are water hardening. In general, tungsten steel magnets are forged, then machined to shape.

Chromium steels

Chromium magnet steels owe their development to the shortage of tungsten during World War I. As shown in the table (see next page), the maximum energy product values of the chromium steels are not as high as those of the tungsten steels. However, magnetic differences between the two materials are not appreciable and the chromium steels have the advantage of being lower in cost.

Compared to carbon steels, the tungsten and chromium steels are more stable after temperature changes and mechanical shock but, nevertheless, they still show some magnetic variations upon aging. Because of their high remanence the tungsten and chromium steels are useful for magnets requiring a small cross-sectional area.

Cobalt steels

The cobalt steels were largely developed in Japan where investigators found that additions of about 35% cobalt greatly increase the coercive force of steels containing 5 to 9% chromium and small amounts of tungsten. In general, an increase in cobalt content is accompanied by an approximately linear increase in coercive force and an increase in residual induction and maximum energy product. Higher coercive forces can be obtained with the cobalt steels than with the carbon, tungsten or chromium steels.

A wide range of magnetic properties can be obtained with the cobalt alloys within suitable composition limits. As shown in the table, manufacturers have established a number of standard alloys, the principal ones containing 3, 9, 17 and 36% cobalt. All of these alloys are martensitic and contain iron-carbon precipitates throughout the crystal lattices as a result of quenching from about 1475 to 1750 F.

Before the Alnico alloys were introduced, the 36% cobalt alloy was the best permanent magnet material known, the maximum

How Permanent Magnets Behave

In order to design a permanent magnet that will perform a given function, the designer must have a basic understanding of how ferromagnetic materials behave. The following description, supplied by Indiana Steel Products Co., explains how the magnetic properties of materials are obtained, and explains their significance.

Understanding the hysteresis loop

Fig 1 shows how the induction in a magnetic material changes as the magnetizing force is varied. When demagnetized material is subjected to a gradually increasing magnetizing force up to H_{max}, the induction in the material increases from zero to B_{max}. If the magnetizing force is then gradually reduced to zero, the induction decreases from B_{max} to B_r on the vertical axis. This B_r value is known as the residual induction.

If the magnetizing force is reversed in direction and increased in value, the induction in the material is further reduced, and it becomes zero when the demagnetizing force reaches a value of H_c, known as the coercive force. A further increase of this negative force causes the induction to reverse direction, becoming $-B_{max}$ at $-H_{max}$. If the magnetizing force is reversed from this point to H_{max}, the change in induction is along curve $-B_{max}$, $-B_r$, B_{max}. This cycle gives the complete hysteresis loop.

Such a curve applies to all magnetic materials, the difference in materials being largely a matter of the values. Materials having a low coercive force are low-energy materials, and those having a high coercive force are high-energy materials. These are commonly known as soft and hard materials, respectively, but the terms low-energy and high-energy are more representative of the characteristics of magnetic materials.

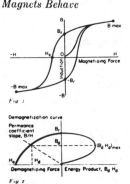

Fig 1

Fig 2

Obtaining the maximum energy product

The section of the hysteresis loop from B_r to H_c is of major interest to designers of permanent magnets. This is known as the demagnetization curve and is shown in Fig 2. At the right of this curve is the conventional energy product curve, which is the product of B and H as taken from the demagnetization curve and plotted against B. The product of B_d and H_d at any point on the demagnetization curve indicates the useful energy produced per unit of volume. In the cgs system where B_d is in gauss and H_d is in oersteds, $B_d H_d/8\pi$ is equal to energy product in ergs per cu cm. The external energy is zero at both B_r and H_c, and reaches a peak value at a point known as the peak energy product, $(B_d H_d)_{max}$. This point represents the maximum external energy that can be produced by a unit volume of a given material. It is a criterion for comparing different permanent magnet materials. Furthermore, in the case of fixed air gap applications the design which causes the magnet to work at the maximum energy point will require the least volume of magnet material.

energy product being 1.0×10^6 gauss-oersted. The alloy is still used today in a limited number of

applications; however, it is relatively expensive for the amount of energy that it produces.

Figure 17-3.   Detail as optional reading

maintaining the same precision in their writings regardless of the audience they are addressing. They would rather remain speechless than be caught describing high polymer thermoplastic fibers as resembling tiny corkscrews hooked together and arranged in parallel, intertwining rows, or referring to an edge dislocation in a crystal as being similar to a scar in the flesh.

But a certain degree of conscious or deliberate inaccuracy is necessary and permissible in order to make our detail palatable and understandable to other than our wise and knowing colleagues. In an editorial in *Science* some months ago, Warren Weaver[2] proposed the idea of "communicative accuracy" to gauge the degree of inaccuracy permitted in scientific writing. He suggested: "A statement may be said to have communicative accuracy, relative to a given audience of readers or hearers, if it fulfills two conditions. First, taking into account what the audience does and does not already know, it must take the audience closer to a correct understanding. . . . Second, its inaccuracies (as judged at a more sophisticated level) must not mislead, must not be of a sort which will block subsequent and further progress toward the truth. Both of these criteria, moreover, are to be applied from the point of view of the audience, not from the more informed and properly more critical point of view of an expert."

HOW TO CONVEY DETAIL

Deciding the kind, amount, and accuracy of the detail is only half the job. We must still help our readers to read, comprehend, and remember the details we have decided to offer them. To tell us how we can do this, we again can refer to what is being taught to our readers in their better reading classes.

One of the important techniques they are taught is to relate detailed facts to the main ideas to better understand and retain the detail. Nila Smith, a leader in the field of reading instruction, puts it this way: "As you read, think of the main idea as a magnet drawing the particles toward it—the 'particles' being the smaller and detailed ideas. Then visualize the main idea together with its cluster of subideas as a unit in itself. In a nutshell, these are the processes which will enable you to grasp and hold in mind a series of minor factual details."

Translating this reading rule into a rule for writing, we can say that each detail must be directly and intimately related and connected to its main subject or idea. This means every detail in its proper place. And the proper place is with the idea it is supporting.

To emphasize the importance of this rule and also to summarize many of the points covered in this article, there is reproduced in Figure 17-4 a small section of a long article in which the author has failed the reader.

In reading this piece of writing, even the brightest readers would have

A   Because of the essential lack of atmosphere at orbiting altitudes, the heat transfer problems of satellite bodies are considerably different from those described for vehicles within the atmosphere. <u>Free molecules which exist at altitudes of 100 miles or higher do not develop convective heating.</u> Thus, the only significant heat transfer process involves radiation

B   heating of the skin due to the solar flux. The thermal problem in this case does not primarily involve the structure of the satellite but the delicate electronic equipment which is the payload. For this reason the concern is with temperature changes from normal involving rather

A   low values.

B   The primary external sources of heat to the satellite include <u>the radiation from the sun and the reflected and emitted radiation from the earth and atmosphere.</u> By far the most important parameters of the structure are <u>the emissivity and absorptivity of the skin.</u> High polish may be used to give a high value of reflectivity and thereby a low value of absorptivity. Another approach involves the plating of a light metal shell with a material of high

C   infrared reflectivity covered by a transparent layer having a hard surface which would be <u>resistant to meteoric dust.</u> A low value of absorptivity generally signifies a low value of emissivity. This condition presents a problem if internal heating is developed by electronic equipment because of the difficulty of radiating the internal heat. An alternate procedure would be to rely on high emissivity and adopt spinning techniques to continually rotate the position of radiation heating to a position of radiation cooling.

B   As the satellite revolves about the earth, it passes from the earth's shadow into sunlight and back into the shadow again in periodic fashion. If the reflectivity and emissivity values of the surface are not favorable, a low heat sink body may develop quite respectable temperature cycles. The peak temperature may exceed 1000°F on the sun face side and the base temperature may drop below -200°F on this same face. The earth face undergoes less drastic swings because it never sees the sun in full incidence of its rays and it always sees a relatively warm earth even while in the shadow region. A spinning body, or a body of high reflectivity, may develop very mild temperature cycles. The Explorer satellite, for example, is a low deflectivity, spinning type, while the I. G. Y. satellite depends on high reflectivity. Possible abrasion by meteoric dust would be expected to be critical for a high reflectivity type, leading potentially to its destruction by melting for the case of a light metal skin. The spinning type should not be as sensitive to such abrasion.

C   From a materials viewpoint it is obvious that a lightweight metal may be expected to serve as satellite material. However, depending on the design, surface treatments are required to ensure either high reflectivity or absorptivity values which are an absolute necessity for the protection of electronic components as well as the body itself.

Figure 17-4.   Confusing organization

difficulty, first, in sorting out the main subjects and ideas, and second, in associating the details with the main ideas. But those willing to take time, after some study would see that there are three major subjects and ideas covered:

A. The thermal problem encountered in satellites. The thermal problem involves protecting the delicate electronic equipment from three sources of radiation heating—the sun, the earth, and the atmosphere.

B. The properties of the materials and surfaces involved in these problems. The satellite skin must have high reflectivity and low absorptivity to protect the payload from radiation heating.

C. The materials and surfaces used to solve the thermal problems. A light-weight material whose surface is treated to provide high reflectivity and low absorptivity, and to resist abrasion from meteoric dust, seems to offer the best solution.

In Figure 17-4 the details on each of the three main subjects (labeled A, B, and C) are scattered throughout the section. Although this piece of writing could profit by considerable revision, its readability would be greatly improved by merely relating the details to the main ideas.

CONCLUSION

Writing, as we have been told many times, is one of our most important tools. Much of what we do has little importance until we can communicate it to others.

As this entire article has tried to show, the reader is the most important consideration whenever we write. We cannot expect him to organize our thoughts and put them in the right order; to sort out the relevant details; to relate the detail to the main ideas; and to translate our vague language into correct ideas. As we have seen, even with the best writing, the reader must work to get our ideas from the printed page. So let's ease his job as much as we can by practicing the rules of good reading when we write.

REFERENCES
1. M. W. Thistle, "Popularizing Science," Science, April 25, 1958.
2. W. Weaver, "Communicative Accuracy," Science, March 7, 1958.

DISCUSSION AND ACTIVITY

1. Like Farkas, Clauser begins by emphasizing the importance of knowing your purpose in writing and your intended audience. How aware of these two consider-ations have you been in your previous writing?
2. How does knowing the rules of good reading help you become a better writer?

3. How can the writer help readers identify the hierarchy of ideas in a piece of writing?

4. What are the problems and solutions of dealing with detailed information?

5. Study carefully Clauser's suggestions for helping readers identify the hierarchy of ideas and follow the details of a piece of writing. Then rewrite something that you have written earlier—for your business or technical writing course or some other course—and put to use what Clauser says.

Selection Eighteen

The Art of Quoting

JACQUES BARZUN AND HENRY F. GRAFF

Jacques Barzun, widely known as a scholar, teacher, and author, was on the faculty at Columbia University for forty-eight years, finally serving as university professor from 1967 until his retirement in 1975. Henry F. Graff is on the faculty at Columbia University. Their article examines the philosophy and mechanics of quoting.

Whether a researcher writes well or ill, he finds himself repeatedly quoting and citing. This is true regardless of his subject. . . .

Quoting other writers and citing the places where their words are to be found are by now such common practices that it is pardonable to look upon the habit as natural, not to say instinctive. It is of course nothing of the kind, but a very sophisticated act, peculiar to a civilization that uses printed books, believes in evidence, and makes a point of assigning credit or blame in a detailed, verifiable way.[1] . . .

THE PHILOSOPHY OF QUOTING

The habit of quoting in nearly every kind of printed and spoken matter and the rules for doing it are quite recent developments in Western culture.

From *The Modern Researcher*, rev. ed., by Jacques Barzun and Henry F. Graff, © 1970 by Harcourt Brace Jovanovich, Inc. Reprinted by permission of the publishers.

[1] The vagaries of quoters and misquoters are studied and illustrated by Paul F. Boller, Jr. in his *Quotemanship: The Use and Abuse of Quotations for Polemical and Other Purposes*, Dallas, 1967.

159

Formerly, the practice was limited to scholars, and was taken as a sign of the unoriginal, timid, pedantic mind. Although Montaigne's *Essays* and Burton's *Anatomy of Melancholy* were admired for their abundance of quaint quotations, most writers preferred to appropriate the knowledge of others and to give it out again in their own words. Emerson, by no means an unscholarly man, expressed a common feeling when he said: "I hate quotations. Tell me what you know." And another scholarly New Englander, of our century, John Jay Chapman, pointed out that what the great quoters seize upon, they alter as they repeat it.[2]

The views of these two American writers should be kept in mind, not as a bar to quotation or as a license to quote inaccurately,[3] but as a reminder that *your* paper, *your* essay, *your* book should be primarily *your* work and *your* words. If you have not made other people's knowledge your own by mixing it with your thoughts and your labor of recomposition, you are not a writer but a compiler; you have not written a report but done a scissors-and-paste job.

And the chief defect of such an evasion of responsibility is that the piece will probably be tedious to read and lacking in force and light. Many writers of master's essays and doctoral dissertations think that what is expected of them is a string of passages from other authors, tied together with: "on this point he said: . . ." and "in reply, he stated: . . ." These are varied with: "Six months later, Thomson declared: . . ." and "Jennings thereupon differed as follows: . . ." The effect is of an unbearable monotony. Every page looks like a bad club-sandwich—thin layers of dry bread barely enclosing large chunks of some heavy solid.

Unfit for handling, the sandwich falls apart, and the reason is easy to see: unless your words and your thought predominate in your work, you lose control of your "story." The six or eight people whom you quote in any given section had different purposes from yours when they wrote, and you cannot make a forward-moving whole out of their disjointed fragments. This fact of experience gives rise to the first principle of the art of quoting: *Quotations are illustrations, not proofs.* The proof of what you say is the whole body of facts and ideas to which you refer, that is, to which you *point.* From time to time you give a *sample* of this evidence to clinch your argument or to avail yourself of a characteristic or felicitous utterance. But it is not the length, depth, or weight of your quotations that convinces your reader.[4]

---

[2] *Lucian, Plato, and Greek Morals*, Boston, 1931, 3–4.

[3] But as H. W. Fowler says in his *Modern English Usage* under "Misquotation":

"The misquoting of phrases that have survived on their own merits out of little-read authors . . . is a very venial offence; and indeed it is almost a pedantry to use the true form instead of so established a wrong one; it would be absurd to demand that no one should ever use a trite quotation without testing its verbal accuracy."

[4] An apparent exception to this rule occurs when you try to prove a point by reproducing documents. The exception is only apparent, because documents that are longer than a couple of pages should be relegated to an appendix and only discussed or quoted from in the text.

Quote seldom; cite often

Two rules of thumb follow from the principle just enunciated: (1) Quotations must be kept short, and (2) they must as far as possible be merged into the text. The form of quoting that we have just used in introducing these two rules—stopping dead, a colon, and a new sentence—is convenient in books of instruction; it is awkward in writing that describes, argues, or narrates. Far better is the form that we use in the next line and that incorporates into your sentence "that portion of the author's original words which [you] could not have put more concisely" without losing accuracy.

Longer quotations than this cannot, of course, be inserted entire into your own sentence, but your words can lead to the brink of the other author's remarks and, *with or without an ushering verb, can make the two speakers produce one effect*, like singers in a duet. Consider a passage from the biography of a famous English trial lawyer, in which the author wants to make use of an important letter:

He was bitterly disappointed when his old friend Clavell Salter was given the first vacancy. "I am told that S. is to be recommended," he wrote to Lord Edmund Talbot. "Well, he is a splendid chap and a great friend of mine of thirty years standing. I think he will tell you he owes much to me in the early days . . ." The letter is that of a bitterly disappointed man, and ends with a prophecy about his future, which came almost exactly true. "Well, I am fifty-nine; if my health lasts, I suppose I can enjoy another ten years hard work at the Bar." Within a few months of ten years after the date of the letter he died, almost in harness. Shortly after this disappointment he was approached as to the writing of his memoirs and he discussed the project and even wrote a few pages. "What will you call the book?" he was asked. "Better call it *The Story of a Failure*," he said sadly, and laid aside his pen.[5]

If instead of this running narrative and commentary, the biographer had used the lazy way of heralding each quoted remark with "he said" or one of its variants, we should have halted and started and halted and started at least four times. Notice that the method of Merged Quotation here recommended has the advantage of preventing the kind of repetition that *Punch* once picked up from the London *Times*:

Land at Freshwater, Isle of Wright, is being prepared as a rocket-motor testing site, the Ministry of Supply said yesterday. "This land is being prepared for the ground testing of rocket motors," a Ministry official explained.
"Clear now?" asked *Punch*.

Whole books have been composed on the system of repetition, especially in graduate schools. When the candidate has collected his material, he "writes it

---

[5] Edward Marjoribanks, *The Life of Sir Edward Marshall Hall*, with an Introduction by the Earl of Birkenhead, London, 1929, 377.

up" by the simple process of (1) announcing what the quotation implies, (2) giving the quotation, and (3) rehashing what has just been said. To the reader this is death in triplicate. To the author, who has to pay for the typing and possibly the printing, it is a great waste. Knowing how to quote might have reduced the bulk of paper and wordage by more than a third; for what we have called the merged quotation is usually docked of head and tail—only the body of it plays its part in *your* presentation.

A final caveat: a researcher must quote only what he has himself read (or heard), and he must weigh carefully his choice of source when variant texts present themselves. It is a shortcut but rarely an advantage to use another writer's quotation as one's own; the risks are not worth the time saved. Experienced editors of learned journals estimate that as many as 15 percent of all footnotes contain errors of one kind or another in the average article of even careful scholars. And the texts of quotations are probably no more accurate.

Misquotations can arise in an infinite number of ways. A newspaper report from London, for example, bears the headline: EDEN SAYS ROOSEVELT WAS RECKLESS. The article itself, reporting on the publication of Eden's memoirs, says that the writer has criticized President Roosevelt for "recklessness" in the conduct of foreign policy vis-à-vis the Soviet Union. [6] This is certainly plausible, even if the expression is rather bold for a former diplomat. But dare one quote him as having written it? No. Even though headline and text agree as to Eden's words, a little research—which means going to the book itself—shows both to be wrong: what Eden wrote was that Roosevelt's opinions of European affairs "were alarming in their cheerful *fecklessness*."[7]

THE MECHANICS OF QUOTATION

Certain forms must, as we said, always be observed in quoting. The modern quotatiousness that has made everybody aware of whose words are which, and that comes out in the announcer's ". . . and I quote . . ." or the casual speaker's "quote . . . unquote"[8] has made us all slaves to a common reporting system. But its mechanics have a spirtual meaning: the system became universal because people saw the value of respecting a man's *ipsissima verba*,

[6] *New York Times*, March 21, 1965.
[7] *The Reckoning: The Memoirs of Anthony Eden, Earl of Avon*, Boston, 1965, 433.
[8] There is no necessity for either of these verbal devices; they are affectation based on the formality of the written word. Moreover, "unquote" is nonsense probably born of confusion with "end quote." See Follett, *Modern American Usage*, "quote, unquote," 270–71.

because they acquired some notion of the influence of context upon meaning, and finally because the recognition of rights in literary property made correct "crediting" necessary.

The most important conventions that rule the quoting of words in print have been exemplified in the last three pages, as well as earlier in this book. We can now set them down in a row, with a few relevant suggestions:

1. A quotation is introduced and closed by double quotation marks. A quotation within a quotation carries single quotation marks, and a third quotation, if required, brings double ones again.

2. The omission of a word, phrase, or sentence is shown by three dots. If that omission follows the end of a sentence, the fourth dot that you will observe in some quoted material is actually the period at the end of the original sentence.

3. If intelligibility requires the addition of a word or short phrase (seldom more), the added words should be enclosed in square brackets.[9] These words will generally be possessive adjectives—for example, [his], [your]—or the definite article [the], or a pair of words expanding a pronoun, such as an "it" in a quotation that would be unintelligible without an antecedent. One replaces this "it" with: [the document], or some such unmistakable substitute. One may also supply a word where ambiguity might result from the lack of a context.

4. The spelling, capitalization, and punctuation of the quoted passage must be faithfully reproduced unless (a) you modernize these features when quoting ancient texts, in which case you state your principles at some convenient point; or (b) you correct an obvious typographical or other error, in which case a footnote is required to draw attention to the change.

5. An extension of the foregoing rule accounts for the familiar tag at the end of a quotation: [My italics.] It is perhaps a shade less obtrusive to use the phrase "Italics added"; in any case, a writer should not make a practice of sprinkling italics over other people's prose. If you choose your quotation with forethought and *set it with care* within your own remarks, it will generally not need the italicizing of words to make its point. If on rereading you think the force of the quotation somehow does not make itself felt, try cutting down the quoted words to those that actually contain the point. Then you will not have to underline them in order to make them stand out.

6. The quoting of titles and the like in the text is more properly called "citing" and this sometimes presents small problems. But note at the outset the second invariable rule of modern scholarship: all book titles are printed

---

[9] In legal scholarship, brackets are put around any capital letter that is supplied by the quoter when he changes the grammatical role of a word. This is rarely necessary in history-writing.

titles of books-underlined
titles of Mag. articles- quotation marks

in italics (underlined in typescript), and all essay titles are enclosed within double quotation marks.[10]

Apart from this convention, a question sometimes arises as to the wording of the title to be cited. Readers feel a certain awkwardness in reading: "This he achieved in his *When Knighthood Was in Flower.*" The remedy for this is simple: insert a cushion word or phrase, for example, "in his popular tale," or "in his next novel," or more flatly: "in his book, *When,* etc." In other words, when the title does not merge easily with your sentence, put in a noun with which the title shall be in apposition.

The penchant for merging is a by-product of the desire to write good sentences. This is so strong in good writers that they cut off the *A* and *The* of a title that they cite in running text: "Motley's *Rise of the Dutch Republic.*" This practice is sensible and long established, but if you follow it, take care that you do not subtly alter the author's intention. He may have been scrupulously exact in putting *A* in front of his subject to show that his conclusions are tentative; for instance: AN *Economic Interpretation of the Constitution* by Charles A. Beard. Conversely, an acknowledged master may at the end of his life bring out THE *Theory of the Cell,* that is, the complete and fully organized view of the subject, which fears no rivals for none exist.

Except for these significant uses, the grammatical role of the definite and indefinite articles can be taken by other words: Samuel Smiles's *Life of George Stephenson, Railway Engineer.* We lose nothing by the omission, but watch again: we fall into gibberish if having properly elided a redundant article in one place we begin to use the decapitated title as if it were complete; do *not* write: "He then published *Life of George Stephenson,* etc."; any more than you would write: "Motley worked nine years on *Rise of the Dutch Republic.*" The use of "a" and "the" in all contexts is a delicate measure of a writer's sensitivity to what he is talking about.

7. There should be firmer rules than there actually are about the *right* to quote. Legal doctrine is on this point indefinite, for all it says is that "fair use" is allowed. "Fair use" covers most scholarly and critical quoting, but it sets no clear limits to the amount that may be quoted without special permission. Hence the demand of most publishers that the author of a manuscript ask permission of the copyright holders for each of his quotations. But do you ask for six words? Obviously not. Some publishers allow passages of 250 to 500 words to be quoted without permission having to be asked. This latitude is sometimes stated on the copyright page: look there for the exact wordage granted free. Remember that an aggregate of, say, 300 words quoted by you here and there requires permission as if it were one long passage.

---

[10] Some periodicals devoted to book reviewing dislike the appearance of frequent italics in their pages and require book titles to be named within quotation marks. This practice does not lessen the importance of the rule or the unanimity with which it is observed in the world of research.

All American university presses allow one another's authors a thousand words on the "no permission needed" principle. For the rest, you must write and ask the publisher, giving the length, opening and closing words, and page references of the passage you want to quote, and also the name of your own publisher. Allow plenty of time before your publication date, for you may have to write to the author or to a third party—the owner of the copyright—to whom the publisher will refer you. Most British authors will express surprise at your wanting a written permission for what they do every day without let or hindrance. In this country it is otherwise, and one must be particularly careful about quoting words from popular songs as well as poetry.

"Credit" should naturally be given for every quotation, but there is no need to be overcome by gratitude: you are keeping an author's thought and fame alive when you quote and name him, and for this he owes you a debt. Some publishers will specify for your acknowledgment a formula of their own, occasionally quite long and in a style reminiscent of funerary inscriptions. They insist that its use is compulsory, but open a current book and you will see that thanks are given in a lump to all copyright owners on an acknowledgment page, or that each is named at the point where his material is used.

In citing and crediting authors, you must, as in all other phases of research and writing, use judgment. Many obvious quotations need no author's name. For instance, if you are so unadventurous that you find yourself quoting "To be or not to be," leave the author in decent obscurity. If you quote for a purely decorative purpose an anecdote that is going the rounds and that you found in your local newspaper, or a trifling news item from the same source, no reference is needed to the place and date (see the extract from *Punch* above). Remember, in other words, that quoting is for illustration and that citing is for possible verification. What is illustrative but unimportant (and in any case not likely to have been garbled or forged) will not need to be verified. It would therefore be pedantry to refer the reader to its source. Like Emerson, he wants you to tell him what you know.

DISCUSSION AND ACTIVITY

1. What is the proper use of quotations?

2. What are the two rules of thumb for using quotations?

3. What are the seven suggestions by Barzun and Graff for using quotations?

4. If you have an earlier piece of writing in which you handled quotations unskillfully, revise those parts containing the quotations.

Suggestions for Further Reading

THE PROCESS OF WRITING

Campbell, H. A. "How to Prepare Technical Reports." *Product Engineering* 25 (October 1954), 154–157.

By determining which of nine categories a report belongs in and by reviewing the needs the subject matter of the report best meets, the writer can choose the proper format (memo, informal report, formal report). The nine categories of reports are (1) receiving and inspection, (2) performance of purchased items, (3) tests of current designs, (4) investigations of materials and processes, (5) investigations of troubles and failures, (6) proposals to apply research, (7) literature review, (8) progress reports, and (9) administrative and special purpose reports.

A report should be easily readable, use standard terminology and symbols, employ high-quality graphics and typography, include the essentials in the introduction and conclusion sections, present data for maximum convenience to the reader, and acknowledge contributions of others who aided in the preparation of the report.

The writer should adapt the following generalized report outline: title page, table of contents, introduction, results and recommendations, literature review, theoretical analysis, apparatus and methods, data and discussion, acknowledgments, references and bibliography, tabular data, charts, curves, and photographs, and appendix.

Dilley, David R. "A Business Manager Looks at Business Writing." *English Journal* L (April 1961), 265–270.

Claiming that "one of the most important ingredients in effective writing is knowing and following the sequence of steps . . . necessary . . . to have a satisfactory end product," Dilley describes ten steps that apply to all types of business writing: (1) determining objectives and requirements, (2) preparing and analyzing the information, (3) making the outline, (4) writing the opening, (5) writing the first draft, (6) cooling off, (7) editing, (8) reviewing the manuscript, (9) critically examining the manuscript, and (10) writing the final draft. In addition, Dilley considers eight tools to be developed by beginning writers.

Klass, Philip J. "How to Prepare Papers the Easy Way." *Aviation Week*, July 2, 1956, 69–71.

Calling the reading of papers at professional meetings "one of the brutal forms of torture devised by civilized man," Klass suggests four principles for improving the effectiveness of papers: (1) determine specific primary objectives of the paper and do not have so many objectives that they cannot be achieved in the amount of time or space allotted; (2) analyze audience to determine what it already knows about the subject matter, what it would like to learn, and what it can profitably use; (3) capture interest quickly by explaining what is new about the information and how the audience can apply or benefit from the information; (4) match the pace of the presentation with the rate at which the audience can receive the information.

Laird, J. Packard. "How to Start to Write a Report." *Product Engineering*, January 23, 1963, 46–47.

To overcome inertia and to get started at writing a report, one should imagine that the completed report is in front of one. The writer should try to visualize who will read the report, how the report will change their thinking, and whether the report is worth the time to prepare. It is important to be discriminating in selecting material, using only information that is related directly to the subject and purpose at hand. The writer should keep the reader in mind at all times and relate information to the reader's needs and interests.

Souther, James W. "Applying Engineering Methods to Report Writing." *Machine Design* 24 (December 1952), 114–118.

The methods of engineering and manufacturing can be applied to writing problems. The eight-step procedure consists of (1) analyzing the problem, (2) planning the treatment of the problem, (3) investigating the problem, (4) designing the product, (5) constructing the product, (6) checking the product, (7) modifying the product, and (8) preparing the final product.

Swift, Marvin H. "Clear Writing Means Clear Thinking Means. . . ." *Harvard Business Review* 51 (January-February 1973), 59–62.

Even routine memos of minor importance need revising until they say exactly what the writer means for them to say. One revises not only to eliminate wordiness but also to achieve precision in one's thinking.

AUDIENCE ANALYSIS AND ADAPTATION

Mills, Gordon H. and John A. Walter. *Technical Writing*, 3d ed. Holt, Rinehart and Winston, New York, 1970.

Readers can be divided into six categories. In order of increasing knowledge about the subject, they are those with (1) little ability to understand complex subjects; (2) considerable education but not in general field of subject; (3) considerable knowledge of general field of subject but not of specific area being reported; (4) considerable knowledge of specific area but little interest in exact topic being reported; (5) considerable knowledge of exact topic but no detailed familiarity with certain aspects of exact topic being reported, and (6) considerable knowledge of exact topic and detailed familiarity with certain aspects of exact topic being reported.

Two other aspects of audience analysis are discussed: (1) the elements in reporting that require adaptation (vocabulary, sentence length and structure, and organization), and (2) the communication situation (who are the users and what are the uses of the report).

Pearsall, Thomas E. *Audience Analysis for Technical Writing.* Glencoe Press, Beverly Hills, Calif., 1969.

Audiences are classified into five levels: (1) the lay person, (2) the executive, (3) the expert, (4) the technician, and (5) the operator. Because each audience will have its own purposes in reading one's writing, one must consider what the reader is going to do with the information one gives him or her. General suggestions for writing to each audience level are given, mostly in terms of how much background information should be provided, what needs defining, what types of sentence structure and what sentence and paragraph length are best, and what kind of illustrations are effective.

GRAPHICS

Arnold, Christian K. "The Construction of Statistical Tables." *IEEE Transactions on Engineering Writing and Speech* EWS-5 (August 1962), 9–14.

Clear tables present statistical information in a concise and orderly manner. Every table should be a self-contained and self-explaining unit and should be as simple as possible. A formal table should be numbered and titled for reference, referred to in the report text, have column captions and line headings, have tabular material centered in each column, and have ruled lines if they increase clarity and enhance appearance.

Beatts, P. M. "Use Your Reader's Eyes." *IEEE Transactions on Engineering Writing and Speech* EWS-2 (January 1959), 6–11.

Illustrations help the writer maintain clarity, brevity, and authority by making the abstract more concrete, the unfamiliar more familiar. The equation, considered "a translation of a statement into a second language," should not be considered an illustration. Excluding the equation, the seven types of illustrations are (1) engineering drawings, (2) tables, (3) flow diagrams, (4) graphs, (5) schematics, (6) instrument traces, and (7) photographs.

Bergwerk, R. J. "Effective Communication of Financial Data." *Journal of Accountancy* 129 (February 1970), 47–54.

Busy readers need to see important information at a glance to determine whether further study of a particular report is required. Two of the best ways to highlight information are to use short tables to compare two or three time periods and to use graphs to show trends over several time periods.

Michener, T. S. "Illustrating the Technical Presentation." *Journal of Chemical Education* 31 (June 1954), 318–323.

A writer should give as much attention to the use of illustrations in his or her report as to the writing of it. He or she should know which illustrations best convey certain information—continuous curve graphs show relationships of variables; bar graphs show periodic data and static values; pie charts show relationships of parts of a whole, and so forth.

In selecting scales for graphs, the writer should plot data roughly on graph paper, then rework with larger or smaller scales to improve appearance, increase legibility, and produce the proportions that will fit the required report format.

Rockett, Frank H. "Planning Illustrations First, Simplifies Writing Later." *IEEE Transactions on Engineering Writing and Speech* EWS-2 (June 1959), 56–61.

The writer should put as much information as possible in tables, diagrams, drawings, and photographs, because professionals can retrieve information from them rapidly. Illustrations overcome the optical limitation of written language, have advantages over words for defining, and convey exact information. Written language should supply the necessary transitions between illustrations.

Part IV

What Are Some Important
Types of Letters and Reports?

*"A reader can receive more information and understand it better if
the information is in an expected and familiar form."*
John Mitchell *

Just as there are standard procedures for producing written documents and
adapting them to particular audiences, there are also standard patterns for
organizing information in these documents. Patterns for conveying certain
information in certain situations have been used, recognized as effective,
and refined into widely used, readily identifiable types. Knowing these
expected patterns will help you communicate to readers who expect a
standard pattern to be applied to a specific communication situation. If
you do not use standard patterns, your writing will likely appear to be
unorganized, incoherent, and unemphatic—even if it is not. At worst,
your writing can bewilder uninformed readers; at best, knowledgeable
readers will mentally rearrange your writing. In any event, deviation from
standard patterns can present obstacles to a reader.

The selections in this section each emphasize a major type of letter or
report. They in no way represent the full range of writing that you may
be called upon to do. But they do represent types of writing that you are
most likely to do. Knowing the purpose and general pattern of each will
be a great help to you. Purpose will help you determine what is relevant.
Pattern will give you the structure in which you can express your ideas.

Almost from the first day of employment, you will not only have to
write letters, but you will also be judged by the letters you write. More
likely than not, you will have to write an application letter and be judged,

---

* From *Handbook of Technical Communication*, p. 1.

at least tentatively, on the merits of that letter before you are hired. J. H. Menning, C. W. Wilkinson, and Peter B. Clarke's "Letters about Employment" examines the gamut of correspondence you may have to write when seeking employment.

Instructions and directions, regardless of the format they appear in, are undoubtedly the most frequent type of business and technical writing. The training of personnel and the manufacturing, inspecting, shipping, maintaining, and repairing of equipment rely on some form of instructions. Forrest G. Allen's "Analyze the 'What' Before You Write the 'How to'" explains the nature of such writing and lays down the fundamental rules to follow.

Few projects of any importance can attract financial assistance or gain institutional or organizational blessing from company officers without the submission of a written proposal as part of the project or work request. Lois DeBakey's "The Persuasive Proposal" explains how to succeed in the art of proposal writing. Proposals can be relatively short and simple or long and complex, but the general approach to writing them remains the same.

The facts and ideas of business, technology, engineering, and science are of little use unless they reach others. No profession or discipline is an exception, because it is through reports, articles, and books that professionals share facts and ideas and contribute to the growth of their professions. Herbert B. Michaelson's "Structure, Content, and Meaning in Technical Manuscripts" analyzes the qualities of a good technical report or paper intended for professional colleagues.

Persons at the top of an institution or organization are starved for information. They are responsible for planning, coordinating, and overseeing activities, but are often far removed from the actual work. Thus, they require a steady diet of all kinds of reports, for without them ignorance and guesswork replaces intelligence. Raymond A. Rogers's "How to Organize a Research Report for Management" suggests how to put together a report for supervisors, managers, and executives.

Institutions and organizations strive to spread the word of their activities, capabilities, and attainments to employees, prospective employees, stockholders, customers, and the business world at large. The corporation annual report has evolved into a specific type of report to do just that. A. M. I. Fiskin's "The Corporation Reports to the Reader" is a sprightly but penetrating examination that describes the main features and the writing of such reports.

Selection Nineteen

Letters about Employment

J. H. MENNING, C. W. WILKINSON, AND PETER B. CLARKE

J. H. Menning was professor of marketing at the University of Alabama, where C. W. Wilkinson is currently professor of behavioral studies. Peter B. Clarke is president of Arcus Company.

WRITING THE PROSPECTING APPLICATION

A salesman, like a fisherman, almost has to have a feeling of success, consciousness, and optimism so that he can think positively. You can hardly force yourself to really try to catch fish or make sales unless you feel that you have an attractive bait or an appealing product or service. Other people (and seemingly even fish) sense quickly how you feel about yourself, and respond accordingly. And since application letters are sales letters in every way— sales letters selling your services—you have to have self-confidence and positive thinking.

With a well-prepared data sheet you will have done a good job of lining up your qualifications, of realizing what you can do, and of deciding on those qualifications which most nearly equip you for efficient performance. You are then in much better shape to write an application letter—a sales letter selling your services.

At times you may want to send a prospecting letter without a data sheet. That's your decision. We don't think it's the better decision; most personnel people prefer to receive a data sheet. Even if you elect not to use one, you'll

Reprinted with permission from *Communicating through Letters and Reports*, Sixth Edition, by J. Menning, C. Wilkinson, P. Clarke (Homewood, Ill.: Richard D. Irwin, Inc., 1976 ©.), pp. 350–374.

write a better letter for having prepared one. Having prepared it, you're throwing good money away if you don't let it work for you.

You're also being very foolish if you fail to capitalize on your investment of time and effort (and maybe even cash) by slavishly following the points and aping the style of another person's application letter. The good "model" application letter doesn't exist—and never will for applicants of average intelligence and above. They realize that *the application letter must be an accurate reflection of the writer's personality as well as aptitudes.* And so they will write their own.

SECURING FAVORABLE ATTENTION

As in sales letters, the infallible way to secure interest in your application letter is to stress your central selling point in writing about serving the reader. Your central selling point may be an ability based on education, experience, or personal qualities or a combination of them. [This applicant] successfully combined all three:

With my college background of undergraduate and graduate work, my teaching experience, and a temperament which helps me to adapt easily to college people and circumstances, I believe I could do a good job as a field representative for your firm.

And after talking with several field representatives about the nature of the work, I know I'd have the added factor in my favor of being very enthusiastic about it.

While I certainly don't know all the reasons why college teachers choose certain textbooks, I have taught enough while completing a master's degree at Alabama to realize that format and price are only minor factors affecting a teacher's decision.

Possibly the most significant realization from my year as a graduate student and instructor is that there is no true "academic" personality—that a successful representative has to be prepared to meet and talk freely and convincingly with a wide range of personalities.

Teaching classes in Economic Problems and Policies, discussing my thesis with committee members both collectively and individually, and talking with staff members about teaching problems (in staff meetings and in bull sessions) have helped me to think on my feet, to have self-assurance when speaking to groups and to individuals, and to adapt myself to varying situations. I've learned to feel at home with all types of college teachers.

The fact that I have studied business at Alabama rather than liberal arts at an Ivy League school may actually make me a better representative, Mr. Dayton—especially if I'm assigned to the South, where I already know the territory. I could serve happily as your representative in any district, however; I've traveled over most of the U.S. (and to Europe and the Far East while in the Navy) and can adapt readily to the fine people and country one finds everywhere.

I believe you'd find me quick to learn; the men I've listed as references on the enclosed data sheet will probably tell you so if you'll write them.

After you've had a chance to verify some of the things I've said about myself in this letter and on the data sheet, will you write me frankly about the possibilities of my working for you?

Possibly I could talk with one of your regional representatives in this area as a preliminary step. And I can plan to go to New York sometime this summer to talk with you further about my successfully representing your firm.

(You may be interested to know that the 22 copies of this letter brought 22 replies within a couple of weeks. Half a dozen of the firms wanted to interview the applicant right away, another half dozen, within a month afterward. The writer had four job offers.)

To get started rapidly and pertinently, one applicant began her letter to the American Red Cross this way:

I can be the versatile type of club director the American Red Cross seeks.

As a result of five years' specialized training in dietetics and institutional management and 10 years' practical experience in meeting and serving people as a volunteer worker in service clubs from New York to Trinidad, from France through Germany, I know the kind of program which will best meet the needs and interests of service men and their families everywhere.

A young man just graduating from college got favorable attention with this:

Because I have had an unusual five-year educational opportunity combining the study of engineering and management, I feel sure of my ability to do efficient work in your industrial engineering department and to steadily increase in usefulness.

I could conduct a time study with a technical knowledge of
the machines concerned or work on the problems of piece wage
rates without losing sight of the highly explosive personnel
situation involved.

A 19-year-old girl with two years of college summarized her outstanding
qualifications in the following well-chosen lead:

As a secretary in your export division I could take your
dictation at a rapid 120 words per minute and transcribe it
accurately in attractive letters and memos at 40 words per
minute—whether it is in English or Spanish.

There's nothing tricky about these openings. They just talk work.

You may be able to capitalize on a trick in some situations—provided it
shows knowledge of job requirements. The young advertising candidate who
mailed a walnut to agencies with the lead "They say advertising is a hard nut
to crack" got results from the message he had enclosed in the walnut. The
young man who, in seeking radio work, wrote his message in the form of a
radio script marked "Approved for Broadcast" and stamped with a facsimile
of the usual log certification indicated above-average knowledge of working
conditions. The secretary who started her letter with a line of shorthand
characters indicated qualifications from the start. The statistical worker who
drew at the top of his letter a line graph showing the Federal Reserve Board
Index of Industrial Production and in the opening lines of his letter commented on the significance of its recent movements certainly had a head start
on other candidates for the job. If you can think of one which is pertinent, in
good taste, and not stereotyped (such as the balance sheet from an accounting
candidate), it may help you. But it is by no means a must and can do you
more harm than good unless you handle it carefully and thoughtfully.

You do need to concentrate on rapidly and naturally establishing your
qualifications with the attitude that you want to put them to work for the
reader in some specific job. Having held out such a promise, you need to back
it up.

SUPPLYING EVIDENCE OF PERFORMANCE ABILITY

Your evidence in an application is simply an interpretation of the highlights of
your data sheet. For persuasiveness, you phrase it in terms of "doing something for you." If you didn't notice how each of the paragraphs two through
seven in the letter beginning on p. 174 gives evidence in support of the
opening promise, go back and read the letter again.

The applicant to the Red Cross whose opening you read in a preceding
passage continued her letter this way:

With the full realization that the Red Cross necessarily operates on an economical basis, I can use my thorough college training in institutional organization as a sound basis for financial management, cost control, personnel management, employee training, and job specification, all of which I know are vital in a well-run Red Cross club.

When it comes to food service, I feel at home in the planning, selection, buying, preparation, and serving of party food for a group of 500 or 1,000 or behind the snack bar of a canteen or in planning the well-balanced meals for the hardworking Red Cross girls who live in the barracks. During my year as assistant dietician at Ward Memorial Hospital in Nashville, I successfully supervised the preparation and serving of from 3,000 to 20,000 meals a day.

Having been an Army wife and lived in many places under varying circumstances, I have learned to use my own initiative in developing the facilities at hand. I've learned to be adaptable, patient, resourceful, and--through grim necessity as a widow--cheerful! I believe in punctuality but am not a clock watcher. And I know from experience that I can direct people without incurring resentment.

I've always enjoyed and participated in the many sports and social activities that are listed on the enclosed data sheet. As a Red Cross director I could help others to share their pleasures too.

The industrial-management applicant followed up his opening like this:

The program I followed at Northwestern University required five years of study because I felt that qualification for work in industrial management should include basic engineering information. The scope of such courses as Business Organization and Cost Accounting were therefore enhanced and expanded by related work in Machine Design and Properties of Engineering Materials.

Three years in the Corps of Engineers of the U.S. Army form the main basis of my experience. A large part of this time I spent as a section officer in a large engineer depot. The knowledge, skills, and experience I gained concerning layout, storage, freight handling, and heavy packaging relate very closely to the problems of factory management in the production of heavy machinery. While working with the problems of shipping bulldozer blades, I was gaining

experience that will aid me in understanding the special
techniques required in handling cotton pickers and tractors.

I've learned how to get my ideas across in business-writing
courses here at Northwestern as well as through being a
reporter for the Daily Northwestern. As a member of the
student governing board and the senior council, I've had
good lessons in cooperation and patience. And despite a
pretty rugged schedule of classes and extracurricular
activities, I've kept myself in good physical condition by
participating on my fraternity's intramural basketball and
football teams.

The enclosed data sheet and inquiries to the men I've listed
will probably give you all the information you want about me
before seeing me, but I shall be glad to furnish any further
particulars you may wish.

And the secretarial applicant to the exporting firm continued (after her
opening) in the following vein, drawing exclusively on her schooling:

In secretarial courses during my two years of study at
Temple College, I've consistently demonstrated my ability to
handle material at these speeds. And as a matter of practice
in my course in conversational Spanish I take down what my
teacher and my classmates say. I have no difficulty
transcribing these notes later.

I learned a good deal about your markets and your clientele
while doing research for a report I submitted this semester
in marketing, "Some Recent Developments in Latin-American
Markets." In the process I became familiar with such
publications as The American Importer, Exporting, and The
Foreign Commerce Yearbook.

I'm neat and conservative in appearance. Early in my life
mother impressed upon me the desirability of a low-pitched
voice and distinct enunciation; probably for that reason my
college speech teacher has been especially interested in
helping me to achieve poise and dignity before a group of
people. On the telephone or in person I could greet your
clients pleasantly and put them at ease.

After I start working, I hope to further my knowledge of the
people and language of Latin America by using my vacation
time for trips to Mexico, Central America, and South
America.

Overcoming deficiencies is a function of the letter, not the data sheet. In
almost any application situation you'll have one or more. In many cases the

wiser course of action is simply not to talk about it! In other cases, if you feel that it is such an important consideration as to merit identification and possibly discussion, embed it in your letter, and endow it with as much positiveness as possible.

The young man wanting to be a publisher's representative had two strikes against him and knew it: he had gone through a commerce school and he was a product of a state university in the South rather than an Ivy League school. Turn back and note how in the fifth paragraph of his letter he met the issue head on and capitalized on it.

The industrial-management applicant had no experience. But did he apologize for it? Not at all! He held out his service experience confidently and showed its relation to the job sought. "Three years in the . . . U.S. Army form the basis of my experience," he wrote—instead of the weak-kneed statement, "The only experience I've had was in the Army," or even worse, "I've had no experience. But I did serve with the Corps of Engineers in the Army."

Probably one of the finest examples we've ever seen of turning an apparent handicap into a virtue is that of a young woman who at first didn't know where to turn when confronted with the necessity for getting a job. After thoughtful analysis of what she had done in college and how it could be used in business, she sent the following letter to a large Chicago mail-order firm. The third paragraph is the epitome of positive thinking.

Isn't one of the significant qualifications of a correspondent in your company the ability to interpret a letter situation in terms of the reader?

Because I believe that I could express an understanding of a situation clearly and imaginatively to your customers (a degree in English from the University of Illinois, an A in Business Communication, and the editorship of my sorority paper suggest that I can), will you allow me to become a trial employee in your correspondence division?

Learning your particular business policies and procedures in writing letters would come quickly, I believe; I am used to following assignments exactly, and I have no previous working experience to unlearn.

I have a good background in writing. I can type 60 words a minute. And the varied extracurricular activities listed on the enclosed data sheet are my best evidence for telling you that I've successfully passed a four-year test of getting along with people.

Will you call me at 876-2401 and name a time when I may come in and talk with you?

It worked! And the same kind of positive approach to any handicap you may have—physical or otherwise—is probably your best way to treat it.

Talking the special language of your reader's business also convinces your reader of your performance ability and helps to overcome any deficiency. In all the examples you've been reading in this analysis, you've probably noticed that each incorporated specific and special references to conditions or products or activities peculiar to the given job. Such references certainly further the impression that you are aware of job requirements and conditions. The would-be publisher's representative referred to books, teachers, college circumstances, and adoptions (the end and aim of that particular job). The industrial-management applicant referred easily and sensibly to two products of the company, tractors and cotton pickers. The applicant to the Red Cross referred to service clubs, canteens and the hardworking Red Cross girls who live in the barracks.

From your research you can readily establish such references. If significant enough information, they may be good choices of talking points for your beginning, as in the following three instances:

```
With the recent improvements on the foot-control
hydraulic-power lift on Farmall tractors and the
construction of a new implement plant at Poplar Bluff,
Missouri, the International Harvester Company of Memphis
will no doubt be selling more farm machinery than ever
before. As a salesman of Farmall tractors and equipment, I
am sure I could help to continue your record of improving
sales.
```

```
The marked increase in General Motors sales for the first
two quarters undoubtedly reflects the favorable public
reception of the new passenger car models and the new
Frigidaire appliances.
```

```
These increased sales plus the increased production as
announced in your annual report also mean more work for your
accounting staff. I can take care of a man-sized share of
this extra work, I believe—and with a minimum of training.
```

```
The regular Saturday night reports your retail dealers
submit show consumer trends which I want to help you
translate into continued Whirlpool leadership—as an analyst
in your sales department.
```

Each of these candidates continued to talk the terminology peculiar to the job. For example, the sales applicant referred knowingly to farmers and farming activities and to the selling activities of making calls, demonstrating, closing, and—probably most important in selling farm machinery—servicing. Such informed references are highly persuasive in any application

LETTERS ABOUT EMPLOYMENT   181

letter because they establish in a desirable way the impression that the writer is well aware of the work conditions and requirements.

You want to show such knowledge, of course. But if you state it in independent clauses (flat facts which the reader probably already knows) you'll sound wooden and dull.

The desirability of *emphasizing qualifications instead of analysis* will be clearer to you through comparing the following original letter and the revision. The original is almost painful in its flat, obvious statements. It also uses so much space stating requirements of the job that it fails to establish qualities of the applicant. The revision eliminates the flatness and preachiness through implication or incidental reference.

ORIGINAL	REVISED
It takes a secretary who is versatile, accurate, reliable and dependable for a firm like the Brown Insurance Company. I realize the importance of your having such a secretary, and I believe I have the necessary qualifications. Having graduated from the University of Alabama with commercial studies as my major, I am familiar with such machines as the adding machine, Mimeograph, and Comptometer. Since my graduation I have been employed as a secretary with the Reynolds Metal Company. This has given me an opportunity to combine my knowledge with experience. Insurance takes a lot of time and patience. A large amount of bookkeeping is required because every penny has to be accounted for. My one year of accounting at the	My year's work as a secretary, four years' thorough college training in commercial studies, and lifetime residence in Tuscumbia should enable me to serve you well as a secretary and further the friendly relations between you and your clients. Whether you want to send a memo to a salesman, a note to a client, or a letter to the home office, I could have it on your desk for signing within a short time. While earning my degree at Alabama, I developed a dictation rate of 100 words per minute and a transcription rate of 45, which I demonstrated daily during my year's work as secretary with the Reynolds Metal Company. To help with the varied kinds of record keeping in a large insurance agency, I can use the knowledge and skills from a year's course in accounting and my study

University will enable me to keep your books neatly and correctly; and if it is necessary for me to work overtime, I am in good physical health to do so.

Since the Brown Insurance Company has many customers in different parts of the country, a large amount of business letters and transactions are carried on. As your secretary, I could take dictation at 100 words a minute and transcribe your letters accurately and neatly at 45 words a minute.

Even though accuracy and speed are important, personality is an important characteristic too. Because of the many kinds of people who are connected with this type of business, it is important to have a secretary who not only can file, take dictation, and type, but who can be a receptionist as well. Since I have lived in Tuscumbia all my life, I will know most of your clients as individuals and can serve them in a friendly manner.

I have enclosed a data sheet for your convenience.

Will you please call me at 374-4726 and tell me when I can talk to you?

of filing systems, office practices, and office machines--all applied during my year of work. You can trust me to compute premiums accurately, send notices on schedule, and devise and turn out special forms when necessary.

I realize that in an insurance agency everyone from the janitor to the bookkeeper affects the feeling of the public and that all must exercise friendliness and tact in any contact with a client. I anticipate the unexpected, and I meet it calmly; so I am prepared to handle a number of duties and to adjust to the demands of a busy, varied work schedule (including overtime work when it's necessary). I would expect to maintain cordial relations with all your customers quite naturally and easily because most of them are the neighbors and friends I've lived around all my life.

Mr. Bills and the other references I've listed on the enclosed data sheet will be glad to confirm my statements that I can work efficiently and cheerfully for you as a secretary who is able and willing to do more than turn out letters. After you've heard from them, please call me at 374-4726 and name a time that I may come in and talk with you.

Although the revision is a little longer, it accomplishes a good deal more: It establishes qualifications in a good lead; it talks the special language of the reader; it establishes more qualifications. It also has a much better work-for-you interpretation. But the major improvement of the revision over the original is that it eliminates the preachy, flat statements (particularly at the beginnings of paragraphs) that made a smart girl sound dull.

ASKING FOR APPROPRIATE ACTION

Whatever action you want your reader to take, identify it as specifically as possible, and ask confidently for it. Ordinarily it is to invite you in for an interview. As a self-respecting human being who has something to offer, you do not need to beg or grovel; but you do need to remember—and to show your realization of the fact—that the reader is under no obligation to see you, that giving you time is doing you a favor, that the time and place of the interview are to be at the reader's convenience, and that you should be grateful for the interview.

The action ending of the sales letter needs to be slightly modified in the application letter, however. You cannot with good grace exert as much pressure. For this reason most employment counselors and employers do not advocate using any reply device (an employer is happy to pay the postage to send a message to a potentially good employee, and writing and mailing a letter are routine actions). But your application action ending still suggests a specific action, tries to minimize the burdensome aspects of that action through careful phrasing, establishes gratitude, and supplies a stimulus to action with a reminder of the contribution the applicant can make to the firm.

You've already seen several action endings in this chapter. But to drive home the point, let's look at the action endings of the four letters with which we started this analysis.

The Red Cross applicant definitely planned a trip to Washington for job-hunting purposes; so she concluded her letter logically and naturally with

When I'm in Washington during the first two weeks in August, I should be grateful for the opportunity to come to your office and discuss further how I may serve in filling your present need for Red Cross club directors. Will you name a convenient time in a letter to me at my Birmingham address?

The industrial-management applicant phrased his ending in this simple fashion:

Please suggest a time when you can conveniently allow me to discuss my qualifications for work in your industrial engineering department.

And the secretarial applicant confidently asked her exporter-reader:

Won't you please call me at 615-5946 and tell me a time when
I may come to your office and show you how well my
preparation will fit into your firm?

The publisher's-representative applicant was in a slightly atypical situation. He couldn't afford to ask directly for an interview in New York because he had neither the money nor the time right then. So he wrote:

After you've had a chance to verify some of the things I've
said about myself in this letter and on the data sheet, will
you write me frankly about the possibilities of my working
for you?

Possibly I could talk with one of your regional
representatives in this area as a preliminary step. And I
can plan to come to New York sometime this summer to talk
with you further about my successfully representing your
firm.

(As it turned out, he flew to New York at the expense of the firms on two occasions within two weeks after sending the letters, but that was the result of further correspondence—and it's certainly not anything to count on!)

Such letters as suggested in the preceding pages and in the checklist for applications won't work miracles. They won't make a poor applicant a good one. They won't ordinarily secure a job; usually they can only open the door for an interview and further negotiations, but that is their purpose. To make yours do all it can, you may want to review the list of suggestions on pp. 190–191.

WRITING THE INVITED APPLICATION

Often a firm makes its personnel needs known (especially for middle- and upper-management positions) by running an ad, by listing with an agency (commercial, where they'll charge you a fee, or governmental like the U.S. Employment service offices and state-government equivalents), or simply by word of mouth. As you probably know, most large companies also list their needs for college-graduate personnel with college placement bureaus and have recruiting personnel who regularly visit campuses scouting for talented young men and women.

These situations (where the prospective employer actually goes out searching for new employees) give you one drawback (you'll have more competition because more people will know about the job) and two advantages in writing a letter: (1) you don't need to generate interest at the beginning (you already have it!); and (2) the ad, agency, or talent scout will give you the job requirements or as a bare minimum identify the job category and principal duties.

Even when you hear of the job through other people, they will usually tell you what you'll be expected to do. So matching up your qualifications with the job requirements is easier in the invited situation than in the prospecting because your source will usually identify requirements in some order indicating their relative importance to the employer.

If you are equally strong on all points of preparation, you have no problem. You simply take up the points in the order listed. But such a happy condition you'll rarely find. Most often your best talking point is not the most significant requirement, and usually you'll be deficient in some way. The solution is to employ the same strategy you did in writing the invited sales letter; tie in your strongest point of preparation with something the reader wants done; take up those points wherein you are weakest in the middle position of the letter and attempt to correlate them with some positive point.

Your analysis of job requirements and compilation of a data sheet are exactly the same procedures as in a prospecting situation. Adaptation is simply easier. And once past the opening, supplying evidence and asking for appropriate action are the same. Since the beginnings in the prospecting and the invited applications do differ somewhat, we need to consider why and to make some suggestions that will help you write good ones.

Whether you learn of the job through an ad, through an agency, or via a third person, your beginning is pretty much the same. The first requirement is that it mention your main qualifications; the second, that it identify the job; the third, that it show a service attitude; and the fourth that it refer to the source of the information (*subordinately* unless it is significant). The reason for naming this fourth function is simply that the reference to the ad, or the bureau, or the person who told you about the job is an automatic attention getter which favorably reinforces the reader's willingness or even eagerness to read your letter. One good sentence can accomplish all four functions and point the trend of the letter.

The opening of the following letter puts emphasis on service through work, clearly identifies the specific kind of work sought, and desirably subordinates the reference to the source. Note that after the opening the letter reads much the same as a prospecting application (indeed, if you omit the lead in the faked address block and the first two lines, it could be a prospecting letter). Note also the adaptation of talking points—the stress on experience rather than on formal training.

```
I'm "sold
on insurance"
```

and I believe I can be the aggressive salesman for whom you advertised in Thursday's Express.

Five years of experience in dealing with people very similar
to your prospects—in addition to technical training in
insurance and salesmanship—would aid me in selling your
low-premium accident policy.

As a pipeliner in Louisiana in 1971 I made friends with the
kind of men to whom I'd be selling your policies. I had a
chance to study people, their hopes and fears and desires
for protection and security, while doing casework for the
Welfare Society in San Antonio in the summer of 1972. And
while working as a waiter both in high school and in college
I learned how to <u>work for</u> and <u>with</u> the public.

The most significant thing I learned was to keep right on
smiling even though dog-tired at the end of my 6-12 p.m.
shift after having been to school most of the day. And I
certainly learned the meaning of perseverance when I had to
go home after midnight and get on the books for the next
day's assignments.

The same perseverance that earned me <u>B</u>'s in Insurance and
Income Protection, Liability Insurance, and Personal
Salesmanship will help me find leads, follow them up,
persuade, and close a sale. I know an insurance man makes
money for himself and his company only when he sticks to a
schedule of calls. But I'm equally aware of the value of
patience and the necessity for repeat calls.

Because I'm friendly and apparently easygoing, your
prospects would like to see me coming. I was elected a
Favorite at Schreiner Institute, and at the University of
Texas I was tapped for Silver Spurs, a service-honorary
organization. Making these many friends has resulted in my
knowing people from all sections of the state.

My build and obvious good health inspire confidence. And
since I'm 24 and single, I am free to travel anywhere at any
time, as well as to work nights.

Dr. Fitzgerald and the other men I've listed on the enclosed
information sheet can help you evaluate me professionally
and personally if you'll write or call them.

I should be grateful for your telling me a convenient time
and place when I may talk with you further about my
qualifications for being the hardworking salesman you want.

Frequently your source—especially an ad—gives you an effective entering
cue and provides you with useful reference phrases throughout the letter.
From the key phrases you can almost reconstruct the ad the young man
answered in the following letter:

Because of my college training in accounting and my work experience, I believe I can be the quick-to-learn junior accountant for whom you advertised in the May <u>Journal</u> <u>of</u> <u>Accountancy</u>.

Having successfully completed down-to-earth studies in tax accounting and auditing while earning my degree in accounting at Alabama, I should be able to catch on to your treatment of these problems quickly.

And while working as assistant ledger clerk for the Grantland Davis firm in Atlanta one semester, I developed a great respect for accuracy as well as an appreciation of the necessity for the conscientious, painstaking labor so essential in public accounting. There, too, I also saw clearly the necessity for absorbing confidential information without divulging it in any manner to others.

My natural aptitude for synthesis and analysis, strengthened by special study of the analysis of financial statements and reinforced with a broad background of economics, law, and statistics, should enable me to handle the recurring tasks of compiling comparative statements of earnings and net worth. And my training in writing reports will help me to tell the story to my seniors as well as to clients.

Realizing that the public accountant must gain the confidence of his clients through long periods of accurate, trustworthy service, I welcome the offer of a long-range advancement program mentioned in your ad. I'm not afraid of hard work. And I enjoy the good health essential in the long, irregular working hours of rush business seasons.

Will you study the diversified list of courses and the description of my internship listed on the attached data sheet?

Note also, please, the wide range of activities I took part in while maintaining an <u>A</u> average. Then will you write the references I've listed as a basis for letting me talk with you further about my qualifications for beginning a career of immediate usefulness to you?

I can start to work any time after graduation on June 4.

A variation of source doesn't affect your procedure—except that you *emphasize a source that would be influential in your getting the job; otherwise, subordinate the source.* If you learn of work through an agency or a third person, the procedure is still the same. Here are some openings bearing out our statement:

Since I have the qualifications necessary for successful
selling that you listed in your recent letter to the dean of
students here at the University of Illinois, I believe I
could serve you well as a salesman.

When I talked with Mr. Hugh Lomer this morning, he assured me
that I am qualified by experience and professional training
for the duties of a field auditor with your firm.

During the four years I worked as a branch-house auditor for
the L. B. Price Mercantile Company to put myself through
school, I became thoroughly familiar with every phase of
accounting work necessary for a branch office of a large
installment concern and with the reports required by the
home office.

I'd certainly like the chance to prove that my education and
personal characteristics parallel the description of the
desirable management trainee that you gave to Dr. Morley,
head of our placement bureau, when you visited the campus
last week.

Two warnings need sounding, however. The *first* is to guard carefully
against the stupid question, the one with the obvious answer. It is usually the
result of asking a question which is made perfectly clear from the ad or the
situation. When a young lady began her application to a legal firm with—

Are you looking for a college-trained secretary who can do
the work in your law office efficiently and accurately and
who is eager to learn law work? If so, I think I can meet
your exacting requirements for a legal secretary.

—she was earnestly trying to highlight the employer's needs. But the reader
had made the answer to her question perfectly clear in the ad! And an
efficient candidate only looked silly in the eyes of this reader.

You don't need to worry about setting out requirements; they are already
clearly established. Even this opening is questionable because the answer is
so obvious:

Wouldn't that junior accountant you advertised for in the
Tribune be more valuable to your firm if she had a sound
understanding of accounting theory and principles and basic
training in industrial accounting?

The reader would probably snort, "More? She wouldn't be valuable if she
didn't!"

The *second* warning is against showing signs of selfish glee over having
discovered a job possibility of your choice. When you read or hear about the
job, you may rightly think, "That's just what I want"—but don't write this

or any variation of it. Resist the impulse and start writing in terms of doing something for the reader: what you can give instead of what you hope to get.

Perhaps a third warning should be sounded against assuming that you don't have much of a selling job to do because the reader is on the asking end. Nothing could be further from the truth. The competition you're up against for an advertised job is keen even in the heyday of prosperity. And because many others will apply, you'll have to write a superior letter to be chosen as one of the final few for interviewing.

In fact, the reader may face such a heap of letters that yours may not even get read. For that reason you may want to do one of several things so that your letter will command attention and thus be selected for reading. Most of these have to do with the physical impression or the mechanics of sending.

A favorite device is sending the letter by special delivery. Few personnel people ever object. If you are in the same town, you can deliver the letter yourself, with the request that it be turned over to the appropriate person.

If you insert the letter in an envelope large enough to accommodate an 8½- by 11-inch page without folding and put a piece of cardboard under it to keep it smooth, the contrast between your letter and all the others that have been folded will call attention to yours.

Cutting out the ad and pasting it neatly at the top of the page may single out your letter for attention. Beginning your message with a faked address block which quotes from the ad is another device. Hanging indention may help to make a rushed reader reach for your letter instead of another. Even appropriate color may cause the employer to read yours rather than another in the stack.

When competition is keen, you'll need to take the time and exert the effort to be sure that your letter is one of the earliest arrivals. This may mean getting up early to get the first edition of the newspaper and having your material in such shape that you can have a complete, well-written letter and data sheet in the hands of the employer hours or even days before less alert candidates get theirs there. Even though you may not get the immediate response you want, your letter (if it is good) becomes better in the eyes of the employer as poorer ones come in through the mail. Remember, too, that people are relieved by the first application that comes in and feel kindly toward it. It relieves the fear of every such advertiser, that maybe no one will answer the ad.

But none of these devices will make much difference if your letter is not written from the viewpoint of contributing to the firm through effective, efficient work.

As you already realize, the items we suggested to you in the prospecting application checklist (p. 190) apply equally when you write an invited letter. Study them again, and review the additional items at the end of that checklist which are peculiar to the invited situation.

PROSPECTING APPLICATION CHECKLIST

1. The prospecting application must generate interest from the start.

    a Establish early your central selling point of education or experience or both, in terms of doing something for the reader. You may also cite your research on the company or the field, or tell a human-interest story; but they postpone the real message.

    b Avoid the preaching or didactic, flat statement.

    c Avoid implying that your up-to-date techniques are better, or telling the reader how to run the business.

    d Make clear early that you are seeking work of a specialized nature, not just any job.

    e Be realistic; talk work and doing, not "forming an association with." Avoid *position, application, vacancy,* and *opportunity.*

    f You need verve and vigor, not stereotypes like "Please consider my application . . . ," "I should like to apply for. . . ."

    g Don't let your biography drown out what you can do now.

    h Don't give the reader an opportunity to shut you off with a negative response.

    i Mere graduation (rather than the preparation back of it) is a poor lead anywhere, especially at first.

    j Eliminate selfish-sounding statements or overtones of them.

2. Interpretation and tone are important from the start.

    a Maintain a consistent, acceptable tone, neither apologizing for what you don't have nor bragging about what you do.

    b For conviction, back up your assertions of ability with specific points of education or experience as evidence.

    c Generalizing and editorializing are out of place: "invaluable," "more than qualified," even "excellent."

    d Avoid needlessly deprecating your good qualifications.

    e Project your education or experience right to the job.

    f Use enough "I's" for naturalness, but avoid monotony.

    g Show the research and thought which have gone into the project. Address the letter to the appropriate individual if at all possible; talk about company operations and trends in the industry; even a deft, tactful reference to a competitor can be a point in your favor.

3. Your education and experience are your conviction elements.

    a Talk about your experience, schooling, or personal characteristics in terms of accomplishing something. For example, you may register for, take, attend, study, receive credit for, pass, learn, or master a course.

    b The emphasis should go on a phase of work connected with the job you're applying for.

c Refer to education as work preparation (in lowercase letters) rather than exact course titles (in capitals and lowercase).

d You need highlights rather than details in the letter.

e But even highlights need to be specific for conviction.

f Your data sheet supplies thorough, detailed coverage. Refer to it incidentally, in a sentence establishing some other significant idea, just before asking the reader to take action.

g A one-page letter may be desirable, but it's more important that you tell all of your story in the most effective way for you.

4. Reflect your personality in both content and style.

a Refer to the more significant personal characteristics affecting job performance, preferably with evidence that you have them.

b Incorporate phrases which reveal your attitude toward work and your understanding of working conditions.

5. Ask for appropriate action in the close.

a Name the action you want; make it specific and plausible.

b Don't beg and don't command; just ask. And avoid the aloof, condescending implications of "You may call me at. . . ." Usually you ask for an appointment to talk about the job.

c Eliminate references to application, interview, position. Use action references to work and the steps in job getting.

d Clearly imply or state that you will be grateful. But "Thank you for . . ." in present tense sounds presumptuous.

e Show success consciousness without presumptuousness.

f A little sales whip-back at the end will help strengthen the impression of what you can contribute.

FOR WRITING INVITED APPLICATIONS

6. When writing an application in response to an ad or at the suggestion of an agency or friend:

a Primary emphasis should be on putting your preparation to work for the reader. But since your reference to the source is an automatic way of securing attention, you should identify it early and emphasize it if it carries an implied recommendation.

b Avoid stating what the reader would infer ("I read your ad").

c Don't ask questions or phrase assumptions which are clear pushovers: "If you are seeking X, Y, and Z, then I'm your man." "Are you looking for an employee with X, Y, and Z? I have X, Y, and Z."

d Postpone salary talk until the interview if you can. If the phrase "State salary required" is included in the description, your reply of "your going rate" or "your usual wage scale" is acceptable to any firm you'd want to work for.

CONTINUING THE CAMPAIGN

Regardless of the results from your application, you have some follow-up work to do.

If you get an invitation to an interview, you know how to handle it. Accept promptly, pleasantly, and directly (if that's your decision). . . . Just remember to continue your job campaign by indicating continuing interest in serving. . . .

If within a reasonable time you do not hear from the person or firm you've applied to, you'd probably better send a follow-up letter indicating continuing interest.

FOLLOW-UP LETTERS

A good salesperson doesn't make one call and drop the matter if that doesn't close the sale. Neither does a sales-minded job applicant. Even if you receive the usual noncommittal letter saying that the firm is glad to have your application and is filing it in case any opening occurs, you need not hesitate to send another letter two, three, or six months after the first one. It should not be another complete application (yours will still be on file): it is just a reminder that you are still interested.

To have a reason for sending a follow-up letter two or three weeks after the original application, some applicants intentionally omit some pertinent but relatively insignificant piece of information in the original.

```
I noticed in rereading my copy of the application I sent you
three weeks ago that I did not list Mr. Frank Regan, manager,
Bell's Supermarket, Anniston, Alabama.

Since I have worked under Mr. Regan's direct supervision for
three summers, he is a particularly good man to tell you
about my work habits and personality. I hope you will write
to him.
```

Such a subterfuge we cannot commend, if for no other reason than that so many other approaches are available to you. One acceptable one is this:

```
I know that many organizations throw away applications over
six months old.

Because that much time has elapsed since I sent you mine
(dated April 15), I want to assure you that I'm still
interested in working for you, in having you keep my record
in your active file, and in hearing from you when you need
someone with my qualifications.
```

Only a lackadaisical applicant would end the letter there, however. Just a few more words could bring the information up to date and perhaps stimulate more interest in the application, like this:

Since graduation I have been doing statistical correlations at the Bureau of Business Research here at the University. I've picked up a few techniques I didn't learn in class, and I've certainly increased my speed on the computer keyboard and calculator.

I still want that job as sales analyst with your firm, however.

Election to an office or an honorary society, an extensive trip that has opened your eyes to bigger and better possibilities of the job, a research paper that has taught you something significant to the job, and certainly another job offer are all avenues of approach for reselling yourself and indicating continuing interest.

THANK-YOU LETTERS

Following an interview, whether the results seem favorable or unfavorable, your note of appreciation is not only a business courtesy; it helps to single you out from other applicants and to show your reader that you have a good sense of human relations.

Even when you and the interviewer have agreed that the job is not for you, you can profitably invest about two minutes writing something like this:

I surely appreciate the time you spent with me last Friday discussing employment opportunities at Monitor and Wagner.

The suggestions you made will help me find my right place in the business world now.

After I get that experience you suggested, I may be knocking at your door again.

When you are interested in the job discussed and feel that you have a good chance, you're plain foolish not to write a letter expressing appreciation and showing that you learned something from the interview.

Your description of the community relations program of Livania has opened new vistas to me, Mr. Lee.

The functions of the public relations department in your company as you described them made me much more aware of the significance and appeal of this work.

As soon as I returned to the campus, I read Mr. Fields's book
that you suggested and the pamphlets describing U.S. Steel's
program.

Many thanks for your suggestions and for the time you took
with me.

I shall be looking forward to hearing the decision about my
application as soon as you can make it.

JOB-ACCEPTANCE LETTERS

When an employer offers you a job and you decide it's the one for you, say so
enthusiastically and happily in a direct letter that keeps you in a favorable
light.

I certainly do want to work with Franklin & Franklin—and I
didn't need a week to think it over, Mr. Bell, although I
appreciate your giving me that much time to come to a
decision.

I've filled out the forms you gave me and enclosed them with
this letter.

Anything else?

Unless you tell me differently, I'll take off two weeks
after graduation. But I'll call you on Friday, June 11, to
get report-to-work instructions for Monday, June 14.

JOB-REFUSAL LETTERS

Sometime in your life you'll have to tell somebody that you don't want what
has been offered. You may feel that it's routine, that it doesn't mean anything
one way or the other to a busy person who interviews many applicants and
has many other people available. Remember, though, that a human being
with pride and ego is going to read the letter. And make yourself think "I
don't want that job *now*," for you may want to reopen negotiations at some
future point.

   To wind up negotiations pleasantly and leave the way open for you, write
a letter with a pleasant buffer of some favorable comment about the company
or the work, some plausible and inoffensive reason, the presentation of the
refusal as positively as you can phrase it (possibly with the statement of
where you are going to work), and an ending expressing good feeling and
appreciation or both. The following letter is a good example:

Meeting you and talking with you about working for Bowen's
was one of the more interesting job contacts I have had.

The opportunity to learn the business from the ground up and to grow with an expanding company is a challenging one, one for which I am grateful.

As I told you, however, I am primarily interested in product research. Since I feel that my abilities will best be utilized in that way, I am going to work for (a company) that has offered me such employment.

I shall certainly continue to watch your company's progress with interest, and I shall look forward to reading or hearing about the results of your prepackaging program.

LETTERS OF RESIGNATION

When you have worked for a firm, you have benefited in some way (in addition to the regular pay you have drawn). Regardless of how you may feel at the time, remember that you can say something complimentary about how things are run, about what you have learned as a result of your experience, or about the people with whom you have associated. By all means, say it! Then announce your plan to leave, giving consideration to the necessity for ample time in which to find a replacement. In some cases no more than two weeks is enough advance notification; sometimes it should be long enough for you to help train the person who will take your place.

Remember, however, that you want to stay in the good graces of the individuals who have assisted you in your career. You will be wise to give ample notification, to give credit where credit is due. The suggestion to "Be kind, courteous, and considerate to the people you pass on the way up the ladder of success; you will likely meet them on your way back down" is good advice to keep in mind when you leave a job.

In many circumstances your resignation can be oral. And in many circumstances it may be better that way. But when you need to write a letter, consider adaptations of the following:

I've certainly learned a great deal about the clothing market from my work as sales analyst at Foley's the past 18 months.

I shall always be grateful to you and the other personnel who have helped me to do the job and to prepare for a more challenging one.

You will perhaps recall that when I had my interviews with you before starting to work, I stressed my interest in working toward a job as a sales coordinator.

Since I now have such an opportunity at Sakowitz, Inc., I am
submitting my resignation. Apparently it will be some time
before such an opening is available for me in this
organization.

I should like to terminate employment in two weeks. But I can
make arrangements to work a little longer if this will help
to train the person who takes my place.

My thanks and good wishes.

Often when another offer comes your way, you'll feel free to discuss the
opportunity with your current employer before making a final decision. Such
a conference has many advantages for both employee and employer. Often a
counteroffer results, to the mutual satisfaction of both, and the job change
doesn't take place. If, despite a counteroffer, you still decide to make the
change, you can resign in good grace with a letter somewhat like this:

Your recent offer is one I appreciate very much, and it made
me give serious thought to continuing at Bowen's.

Let me say again how much I have appreciated the
cooperation, the friendliness, and helpfulness of everyone
with whom I've been associated here.

After considerably more evaluation, however, I believe I can
make a greater contribution and be a more successful
business manager by accepting the position offered me by
Lowen's.

I hope that I can leave with your approval by (specific
date); I feel sure that all my current projects will be
completed by that time.

You'll hear from me from time to time—if for no other reason
than that I'll be interested in how the new credit union
works out.

But I'll always want to know how things are going for Bowen's
and the many friends I've made here.

When appropriate, a possible talking point is the suggestion of a successor
to you; often this is a big help. A constructive suggestion, phrased positively,
implies your continuing interest in the organization.

Letters of resignation written by college students who resign after having
agreed to work for someone but before actually reporting for work are
something we take up with reluctance. Many personnel people regard them
as breaches of contract. Certainly a practice of sliding out from under such
agreements will soon give you a black eye employmentwise.

We would urge you to give serious thought before definitely accepting a job offer. Don't make the mistake of grabbing the first job offered you, only to have something infinitely more to your liking come along later. We'd further urge you never to let yourself get caught in the position of being committed to two employers at the same time. If you have agreed to go to work for a firm and then you have a later offer which you want to accept, do not accept it until you are released from the first contract. To the second potential employer, reply in some vein like this:

I certainly would like to accept your offer to come with your firm. As attractive as your proposal is, however, I must delay accepting it until I can secure a release from the Jenkins firm in Blankville. After my interview with you, I accepted this position, which at the time appeared to be the most promising available.

Can you allow me enough time to write the Jenkins personnel manager, explaining my reasons and requesting a release? (I can give him the names of two friends who might be suitable replacements.)

This shouldn't take longer than a week to settle. I appreciate your offer, regardless of how things work out.

If necessary, phone the second potential employer, explain frankly, and get approval to wait. But for your own protection, do it *before* writing a letter like the following:

As you know, I am now planning to report to work as an executive trainee shortly after the first of June.

Before I made this agreement with you, I had talked with a representative of the Larkin organization in Sometown concerning possibilities of my working there as an analyst in the quality control division, which is the kind of work I have specifically trained for and know I want to do.

I believe I'd be a better adjusted and qualified employee in the Larkin job. That is the main reason I ask that you release me from my commitment with you. The fact that Sometown is a considerably larger city and that the starting salary is somewhat larger are only secondary considerations.

No doubt you have other people you can call on to take my place, but you may be interested to know that Don M. Jones and Peter Lawson are interested in the Jenkins program. You can get portfolios on both of them through the placement bureau here at school.

Since the Larkin people have agreed to postpone a decision
until I have heard from you, I should appreciate a quick
reply.

You can rest assured that I shall keep my word with you and
that if your answer is no, I shall report to work as promised
and do all I can to be an efficient, cooperative, and
cheerful employee.

Only a Simon Legree would say no to the foregoing letter. If the company
releases you, you'd then write the appropriate acceptance letter to the second
firm; but you should, as a matter of business courtesy, write a short thank-
you letter to the first company.

TWO USEFUL MODIFICATIONS OF APPLICATIONS

The following two letter possibilities for helping you get the job of your
choice are *not printed here with the implication that they will take the place of the
complete sales presentation* we have suggested to you. Because they may help
you sometimes, we simply remind you of them.

THE JOB-ANTICIPATING LETTER

Most personnel people are willing to give advice. And most of them are
pleased with a show of interest in their companies and evidence of long-range
planning on the part of a student. Several of our students have had successful
results from letters like the following, sent in the junior year of college:

A course in the operation of business machines under Mrs.
Lora Osmus in the Statistics Department of Alabama gave me
skill in their operation and showed me the tremendous
possibilities of Burrows equipment for business use.

After comparing Burrows and ABL equipment that was on
exhibit on Commerce Day and talking with the Burrows
representative in charge of your display, I am coming to you
directly and frankly for some help.

Since I have completed practically all of the courses
required for the B.S. in commerce, I am free to elect
practically all courses I shall study next year before June
graduation. On the attached sheet I've listed the courses
I've completed and those I'm contemplating. Will you please

rank the ones you consider most beneficial for a prospective
Burrows representative?

Naturally, I will regard your suggestions as off-the-cuff
assistance that implies no commitment. I'm just trying to
equip myself as well as I can to meet the competition for the
first available job with your company after I graduate.

I shall be most grateful for your comments.

THE TELESCOPED APPLICATION INQUIRY

We realize that good applications take time. They're worth the time, however.

But we also know that sometime, somewhere, you may need to send some
inquiries in a hurry and simply cannot write a complete one. You may be
able to make profitable use of the services of your college placement bureau in
a letter, as one young man did. He was too busy writing a thesis and sitting
for graduate examinations to prepare a thorough application. He sent the
following request and a reply card to six firms:

With completion of an M.S. degree in accounting at the
University of Alabama and two years of retail merchandise
accounting experience, I believe I could make you a good
accountant with a minimum of training—and be able to
advance more rapidly than the majority of accountants you
could hire.

I am not just an accountant: A well-rounded background of
finance, transportation, economics, and other related
subjects will enable me, in time, to do managerial work as
well.

May I have the Placement Bureau here at the University send
you a transcript of my college record together with a
detailed record of my experience, faculty rating statements,
and names and addresses of former employers?

I shall be happy to furnish any additional information you
may want and to be available for an interview at your
convenience later if you will check and return the enclosed
card.

He received replies from all six firms, it's true. But only one resulted in an
interview.

This may be a stopgap measure sometime. But this young man's experience simply reconfirms the fact that an applicant must tell a complete story if
he expects to get a show of effective interest. . . .

DISCUSSION AND ACTIVITY

1. The authors advise you to "talk work" in your application letters. The key to giving evidence of your qualifications for a particular job is to start thinking about the specific skills you have obtained through your education, training, and experience. A prospective employer learns far more about your skills when you point out that you can type 80 words a minute or that you can do silver electroplating or that you have conducted a study of the economic impact of extending Interstate 64 through eastern Kentucky than when you simply say that you have had six semester hours of typing, or Ind Arts 106, or Econ 399. Choose a course in your major field and list the specific skills you have acquired as a result of completing that course.

2. Although the authors' discussion of the résumé is not included in this selection, you will need to know what information about yourself a prospective employer wants in résumé form. Collect examples of application forms from companies, study the kinds of information asked for, and discuss what information is asked for most often.

3. One of the first steps you take in conducting a job search is to find out what kinds of work are available for persons with your qualifications. Conduct a survey of the various kinds of work in your major field that you would qualify for upon graduation or certification. Look for printed material under the title of your planned occupation (and under closely related titles) in various periodical indexes, the card catalogue, the *Occupational Outlook Handbook*, *Occupational Literature*, *The Macmillan Job Guide to American Corporations*, and other sources recommended by your adviser or librarian. Write a report using main headings similar to these:

- history of occupation
- importance of the occupation and its relation to society
- duties, conditions of work, and typical places of employment
- qualifications for entering the occupation
- preparation and methods of entering the occupation
- earnings and advancement

Selection Twenty

Analyze the "What" Before You Write the "How to"

FORREST G. ALLEN

> *Forrest G. Allen, a retired United States Air Force colonel living in Orange, California, now serves as a consultant to management. Allen's article, which is based on methods he developed when serving as director of the Air Corps Communication/Electronics School, explains nine steps you should master before you write job instructions.*

Learning how to do something new can be either an exhilarating or a frustrating experience. It depends on how clear and complete the instructions are. Many people with know-how know so much that they are confused over how much detail to write about. Either they write too little, because they assume "everyone knows that," or too much, because "this is nice to know."

No matter how simple or how complex the job you wish to write about, your writing task will be made easier if you begin by analyzing the job for its "teaching content." This content is nothing more than the essential facts needed to perform the job successfully. By following nine easy steps you will be able to ferret out all of the essential details to include in your instructions.

The analytical method I am about to describe was sponsored some years ago by the U.S. Office of Education, for use in writing vocational training materials. As you work your way through the steps of the analysis, keep in mind this dictum: Get the *essential* facts, get *all* the facts, and get *nothing but* the facts.

MAKE A JOB ANALYSIS WORK SHEET

This is a large sheet of paper (I use one measuring 22 in. wide by 17 in. deep), with rules and headings as in the illustration (Figure 20-1). The two boxes in the first column will be used for almost every job you write about. The boxes in the second and third columns may not all apply to every job you analyze, but the headings in these boxes should be used as a checklist of things to consider.

When you fill in the boxes you may use a pencil, or you may prefer to type the information on slips of paper that can be pasted in the appropriate boxes of the analysis sheet. If you need more space, write the overflow on a separate sheet of paper on which you have written the proper heading. When you do this, be sure to make an appropriate note on your analysis sheet.

Next, fill in the *job title*. This will usually be a "How to do it" type of title, like "How to determine efficiency through work sampling," or "How to tune a motorcycle engine." A *job number* might be needed when a series of related jobs is involved in a course of instruction, as in plant training programs, or in industrial arts education.

The *job objective* is a brief statement of what the reader should be able to do when he finishes reading the instructions. In many cases the *job title* itself will define the objective. But in formal training programs the *job objective* should describe the quantitative and qualitative result to be achieved. For example: "Ability to inspect 50 pieces an hour, making not more than two errors."

LIST THE STEPS OF PROCEDURE

List each step of the job procedure in the best logical order, using the active imperative form to describe the steps as briefly as possible. Begin each step description with an active verb. For example, the steps in changing a flat tire are:

1. Open trunk
2. Get jack, lug wrench, and screwdriver
3. Get spare tire
4. Position jack and lift car
5. Remove hubcap
6. Loosen and remove lug nuts
7. Remove flat tire
8. Mount spare tire on lug screws . . . and so forth.

Detailed instructions for each step aren't needed at this point. But keep them clearly in mind as you carry out each step of the job analysis.

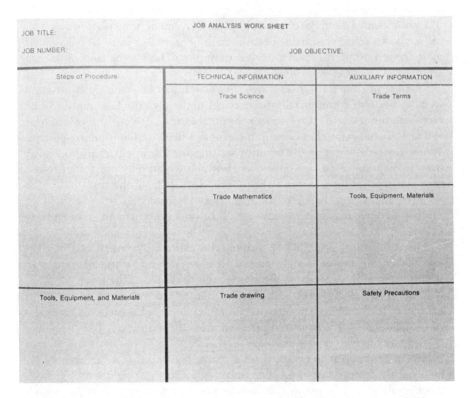

Figure 20-1.   This job analysis work sheet should be filled out as completely as possible

LIST THE TOOLS, EQUIPMENT, AND MATERIALS NEEDED

Use the terminology that will be most familiar to your readers, and include specific information about quantities, types, sizes and (if needed) manufacturer's types and model numbers. This completes the listing of basic information needed for analyzing the job for its "teaching content."

Now you should consider the headings in the second column of the analysis sheet—the one headed *technical information*. In this case, technical information is the essential facts of science, mathematics and drawing that the reader will need to perform the job successfully.

WHAT SCIENTIFIC INFORMATION IS NEEDED?

You don't have to be a scientist to think about the scientific facts on which the success of the job depends. And when you finally get down to writing the final instructions, you probably wouldn't even mention the word "science." The *trade science* box in the chart is simply a convenient filing place where you can jot down reminders about certain scientific facts that must be considered to do the job properly. Here are a few examples of how some facts of science are used in doing various jobs:

Spatters of lime mortar on a new brick wall can be removed by using a weak solution of commercial muriatic acid to dissolve the lime mortar. The fact that muriatic acid (also known as hydrochloric acid) dissolves lime mortar is a fact of chemistry that is used in cleaning a brick wall of mortar spatters. When you finally get to the point of writing the job instructions you don't have to mention chemistry or science, but you would tell your reader to use muriatic acid solution to clean the mortar spatters from the wall.

"Friction creates heat" is another fact of science (from the field of mechanics) that makes it necessary to lubricate bearings and some cutting tools.

The mechanical principle of torque (also called a "moment of force") is used when we loosen a tight nut with a wrench. The use of torque to multiply force is a special application of the principle of the lever.

Finally, the electro-chemical phenomenon of electrolysis is an insidious source of trouble in poorly designed plumbing installations where, for example, copper and galvanized iron pipes are directly connected. The pipes will corrode due to electrolytic action.

If you search deeply enough you will probably find that some "science" is involved in almost every job. The only reason in thinking about "science" when making a job analysis is to identify the technical facts on which successful job performance depends. Ask yourself if the reader *could* do the job properly if he didn't know about the technical fact. Any superfluous or trivial information that you include in your job instructions will only confuse and distract your reader—it will make him wonder what the connection is between the information given and doing the job.

WHAT KNOWLEDGE OF MATHEMATICS?

For the purpose of job analysis, mathematics means anything involving numbers, measurements, or calculations and the accuracy necessary in doing the job properly. For each step of the job procedure, ask yourself questions like:

1. Are any linear measurements necessary, and to what degree of accuracy? What system of measure is used—English or Metric?

2. Is it necessary to locate numbers in a table? Is interpolation necessary?

3. Is it necessary to read and derive numerical data for use in making graphs

or curves? Conversely, is it necessary to read numbers from a graph—and to what degree of accuracy?

4. Are any calculations needed, and how accurate must they be?

5. Does the job require the use of any formulas?

6. Are calibration charts used in doing the job? Is it necessary to interpolate?

7. Is it necessary to read numbers from any meter scales or other devices having scales? What degree of accuracy is needed?

8. Is it necessary to set any dials or controls that require a numerical setting, and how accurately must this be done?

9. Must numerical data be converted from one set of units to another (for example, converting from feet per second to miles per hour)?

WHAT KNOWLEDGE OF DRAWING IS REQUIRED?

To do the job successfully, is it necessary to do any of these things?

1. Make or interpret freehand sketches

2. Make or read drawings and layouts to scale

3. Read blueprints, or similar engineering or architectural type drawings

4. Make graphs or charts

5. Read or make any kind of schematic diagrams (representing, for example, electrical and electronic circuits, plumbing hookups, equipment layouts and connections and the like)

6. Use or make templates or patterns

7. Visualize the appearance of three-dimensional objects from two-dimensional drawings or sketches

8. Make or read perspective drawings

This completes the analysis of a job for its technical information content. Next, you should analyze each step of the job procedure for any auxiliary information that may be needed to do the job successfully. The auxiliary information, listed in the last column of the analysis sheet, involves trade terms; knowledge of tools, equipment, and materials; and safety precautions.

WHAT KNOWLEDGE OF TRADE TERMS IS NEEDED?

Every occupation has its special terminology. The trade terms of an occupation include all occupational terms, nomenclature, jargon, nicknames, acronyms and abbreviations that may be unfamiliar to the average reader.

The careless use by specialists-turned-writers of the esoteric language of their occupations can cause untold confusion in the minds of their readers. During the job analysis make a special effort to identify any unfamiliar trade terms.

When you get to the actual writing phase, each of these terms must be clearly defined. And to avoid semantic confusion, use the terms correctly and consistently. In general, use the language and terminology that will be most easily understood by your readers.

WHAT KNOWLEDGE OF TOOLS, EQUIPMENT, AND MATERIALS IS NEEDED?

Tools and equipment: To perform some jobs properly, it may be necessary for the reader to know the names, types, nomenclature, purposes, properties, and limitations of certain items of tools and equipment used on a job. But any safety precautions in using tools and equipment should be covered under the *safety precautions* box of the analysis sheet.

Materials: To do a job properly often requires knowledge of the kinds and properties of various materials, and the ability to recognize and select materials that are most suitable for the job. This also applies to ingredients used in chemical and cooking mixtures, and to other mixtures used in accomplishing a job.

When materials must be cut or machined or formed, knowledge of their unique working properties is often essential in doing a job properly.

In some cases, your readers may need to know the trade names by which certain materials are known, as well as where the materials can be obtained.

WHAT KNOWLEDGE OF SAFETY PRECAUTIONS IS REQUIRED?

Suitable safety warnings and precautions must be given to prevent (1) injury to the person doing the job, (2) injury to others, (3) damage to tools, equipment and property, and (4) ruining the job itself, or wasting materials.

Review each step of the job procedure to identify places where safety warnings and precautions will be needed. Your most important responsibility in making the job analysis is to decide when to warn your reader against the possibility of accidents, and to tell him how to avoid the accidents.

According to the National Safety Council, the yearly accident rate among do-it-yourselfers is from 650,000 to 750,000 *disability accidents*.

All you have to do now is perform a job analysis and to incorporate the essential teaching elements in your outline wherever logic dictates. After that, writing the job instructions should be much simpler than if you had not made the job analysis.

DISCUSSION AND ACTIVITY

1. Perhaps no other kind of business and technical writing depends so much on correctly identifying the intended reader than does instructional writing. As Allen says: "Many people with know-how know so much that they are confused over how much detail to write about. Either they write too little . . . or too much." Study a set of instructions and determine what information the reader must already possess to follow the instructions.

2. In commenting on the problem of ambiguity in instructional writing, Warren H. Deck states that "a reader must get only one meaning—the correct meaning" from the information. Study a set of instructions, determine whether it contains statements that could be misunderstood, and write a report analyzing the ambiguities (explaining how the statements could be misunderstood) and recommending revisions that could clarify the statements.

Selection Twenty-One

The Persuasive Proposal

LOIS DEBAKEY

Lois DeBakey is professor of scientific communication at Baylor College of Medicine and Tulane University School of Medicine. DeBakey's article explains the skills needed by the investigator who must write a proposal for research: clear delineation of the problem to be researched, careful preparation and outlining, organization of the first draft, and critical revision of the proposal.

"No man, but a blockhead," said Samuel Johnson, "ever wrote except for money." By that criterion, the grant applicant is no blockhead: he writes strictly for money. How can he minimize the likelihood of a rejection by the granting agency? There are safeguards, and many of them are inherent in the way he writes his application. Even those who deprecate a serious interest in language and writing as the concern of "pedants" recognize the utilitarian value of competent self-expression in a grant application. The ability to write persuasively, in fact, is crucial to a successful application, since it is the sole medium of communication available.

The retrenchment of research funds during the past few years has heightened the competition for grants. Gone are the days when research grants went begging. Only about 50 to 60 percent of applications to the National Institutes of Health are now approved by the reviewing committees, and these applications must then survive a strict priority valuation before actually being funded.

From *Journal of Technical Writing and Communication*, Volume 6, Number 1, "The Persuasive Proposal" by Lois DeBakey. Copyright © 1976 Baywood Publishing Company, Inc. This article is included in the forthcoming *New Directions in Technical Writing and Communication*, to be published by Baywood.

We are all salesmen, whether we acknowledge it or not. Even though we do not make our livelihood selling automobiles, household appliances, food, clothing, real estate, advertisements, or other tangible products, effective salesmanship is the basis of most successful human endeavors. We are all daily involved in selling our appearances, our personalities, our characters, our talents, and our ideas. Certainly, the grant applicant is engaged in selling his ideas and his ability to carry out the activity described in his proposal.

Your goal as a grant applicant is to submit a *persuasive proposal*—one that convinces the reviewer(s) to approve the investment of the funds for which you are petitioning (Figure 21-1). You must, then, define yourself verbally as a good risk—as one who is capable of creative thinking, sound reasoning, and productive investigation. You may, indeed, be capable of all these things, but if you cannot, by the selection and arrangement of words on paper, convince the reader that you have these qualities, you may never receive the financial support necessary to demonstrate your skills, for most application referees judge the anticipated value of the proposed research on the basis of the care and precision with which the application was prepared. Careless words have often robbed applicants of the support they seek.

How can you enhance the persuasive force of your proposal? As in report writing, you will need to anticipate what your reading audience will expect from your application. If the sole evaluator is to be the director of your laboratory, you will write the application differently from one to be read by foundation reviewers who may have no special knowledge of the research methods you plan to use. Knowing your audience will help you decide what to include and how much to elaborate. You should therefore find out all you can about the agency or foundation to which you plan to apply: what its purpose and scope of research interests are, and how its applications are evaluated. Some grantors have special application forms, whereas others merely outline broadly what the application should include. Be sure that you have the latest guidelines and forms for applications, since obsolete information can lead you astray. The more you know about the grantor, its policies and purposes, the more likely you are to fulfill its criteria.

Naturally, you will want to select the most suitable agency for your field of interest. Most libraries contain a directory of granting foundations and agencies, with information about the scope of interest of each. The annual reports of agencies also usually list research projects being supported, but if these are not available, you may write directly to the grantor for descriptive information. Artificially adapting your purpose to satisfy the objectives of an agency will suggest a kind of intellectual dishonesty that will work against you.

Because the reviewing process for applications generally takes from six months to a year, you will be wise to begin writing your application well in advance of the projected date to begin the research. Plan on at least two

Figure 21-1.   The money man

months to prepare the application; anyone who has written a proposal recognizes the conservativeness of this estimate. Meeting the stated deadline is imperative.

Preparation of a grant proposal, an essential component of the investigator-educator's life today, need not be as painful as many applicants consider it. Yet even Nobelists like Szent-Gyorgyi[1] approach the task with distaste; grant applications, he admitted, agonized his entire scientific life. I am convinced that faulty methods used for instruction in language and logic are largely responsible for the anguish that most people experience when faced with the task of preparing a formal written or oral presentation. The dull, negative, impractical approach to English grammar and composition adopted by teachers in the early grades has produced generations of verbally handicapped or inhibited graduates.[2] These inhibitions can, however, be released by the application of a few basic principles of effective self-expression.

These principles apply equally to the preparation of a grant proposal and to that of a scientific or technical report, despite several differences in the two forms. The temporal relation between the writing and the research, for example, differs. You write the *proposal before* you know the results and the *report after* (Figure 21-2). The proposal is a forecast; the final report a retrospection. In the proposal, you need to convince the evaluator that you have the critical faculties to recognize a gap in knowledge or other need and the creative imagination to devise a way of filling that gap or need. The successful research applicant is the one who can reduce the contradictions, the illogicalities, and the ambiguities in uncharted regions to predictable

Application ➤ Grant ➤ Report

Figure 21-2.   Temporal relations of grant application, research project, and final report

phenomena expressed in simple, cogent terms. When you write the final report, on the other hand, you have the results before you and so can write factually rather than anticipatorily. In both the application and the report, however, you must persuade the reader that your thinking is clear and your reasoning sound. [2] The difference in time of composition does not, therefore, affect the essential criteria of composition:

Content	*Form*
Worthiness of subject	Clarity
Propriety for the reading audience	Conciseness
Clarity of purpose	Continuity
Accuracy	Transition
Unity of thesis	Consistency
Orderly organization	Simplicity
Coherent development	Readability

PRELIMINARY PLANNING

A frequent, and grievous, error among novices is to begin writing too soon. The result is a diffuse, rambling, confused proposal that lays bare the incertitude and chaotic thinking of the applicant. An obscurely defined, vaguely described, sloppily developed project is bound to suggest murky thinking and is likely to portend haphazard research. Before you write the first word of the proposal, make sure that the concept you plan to examine has some point of originality, that it is well defined in your own mind, and that it is worthy of support. To satisfy these three criteria, you must have a good grasp of current knowledge on the subject, which means that your preliminary bibliographic research must have been thorough. If your application betrays an ignorance of certain vital aspects of the subject, you will be judged poorly qualified to perform the work, and the funds you seek will be refused. Writing down certain ideas as you go along in your preliminary planning often helps crystallize important points in your own mind. If you are later able to use some of this material in the proposal, so much the better.

When you have carefully delineated the purpose and scope of your study, have a good understanding of its relation to present knowledge on the subject, have planned your method of attack, and feel reasonably comfortable that it will yield desirable information, then and only then should you embark on your writing project.

PRELIMINARY AIDS

First, write down on paper the *precise question* or problem you plan to study. Make sure it is based on a sound premise. Then write the possible or *expected answer(s)* or solution(s). Seeing these two ends of your research project on paper will fix the limits of your study firmly in mind. Next write the *title*. Make it accurate, clear, succinct, and provocative. You will, of course, want to reexamine the title later, after you have completed the proposal, but writing it now will help circumscribe the subject in your mind. An informative title is indispensable; select key words carefully to highlight the essence of the subject. To abbreviate a long title, such as:

> Some Theoretical Foundations and Some Methodological
> Considerations for Thesaurus-Construction

you can often use a broad main title, indicating the primary disciplinary category, followed by a more limiting subtitle, such as:

> Thesaurus-Construction: Theory and Method

Titles such as the following cannot be expected to excite the reviewer:

> Investigations into Implicit Speech Inferred from Response
> Latencies in Same-Different Decisions

OUTLINE   With the question, the possible answers, and the title before you, and with the purposes and goals of the granting agency clearly in mind, prepare an outline of the proposal, to insure unity of thesis, completeness of content, orderly organization, and coherent development of ideas. Even if the granting agency provides a format for the application, you will still need to outline your material in order to develop your central thesis cogently. The outline is a skeleton or frame on which to build the proposal; you should use it as a guide, but be its master, not its slave. Feel free to modify it any time you recognize a deficiency or error in it. Because it is merely a preliminary design or pattern, it need not be excessively detailed; in fact, minutiae are often distracting and confusing. A sentence outline, however, is better than a jot outline because it requires you to organize your material in the same grammatical unit that you will use in the actual paper. Once you have made your outline, you can adapt it to any particular application format.

SYNOPSIS   Next, write a synopsis of the proposal, in which you succinctly define the purpose, describe the methods of procedure and their rationale, summarize the expected new knowledge to be gained, and indicate its anticipated significance. The agency's guidelines usually specify the length of the summary, but rarely is more than one page necessary. Since most agencies require a summary of the proposal, the synopsis has a utilitarian purpose beyond encouraging conciseness and giving a bird's eye view of the proposal as a whole. You will, of course, need to reexamine the synopsis after you have completed your full proposal, but writing it at this point will help you put your ideas in proper focus and balance. Use of key terms in the summary will help the reader focus immediately on the salient features of the project. The synopsis should be comprehensible to the nonspecialist.

THE PROPOSAL

With the preliminary steps completed, you are ready to draft the proposal proper. Write the first draft with attention exclusively to orderly sequence and without consideration for grammatical or rhetorical perfection. Stopping to examine each word, phrase, or clause will shift your attention from thought to form and will interrupt the rational flow of ideas. You can always refine your composition later, but if the presentation of your ideas is disorderly, major excision and reconstruction will be required to repair the damage. Use succinct language and assign to appendices any auxiliary information, to prevent cluttering the text with excessive detail.

INTRODUCTION

In any unit of exposition, be it a sentence, a paragraph, or a full composition, two of the most important positions are the first and the last. The introduction, therefore, should be prepared with special care, for if you fail to engage the reviewer's interest in the beginning, you may lose it altogether. Take a lesson from the prosperous house-to-house salesman who attributed his phenomenal success to the first five words he uttered whenever a woman opened the door: "Miss, is your mother in?" That salesman recognized the importance to successful communication of knowing your audience and making your introduction attractive; he engaged the interest of his audience immediately, and he encouraged receptivity in his opening words.

Even if the reader is a reviewer who is compelled to proceed beyond a dull introduction, you risk putting him in an inhospitable mood. To avoid such nonreceptivity, make your introduction straightforward, sharp, and provocative. If you are long-winded in the beginning, the reader has reason to conclude that you are incapable of sweeping away the irrelevancies and the extraneous from the subject under consideration, and that you are not therefore likely to design and conduct an efficient, fruitful program of study. State the problem and the aims clearly and promptly rather than make the reader traverse several pages before he finds out what your central point is. Justify your study of the problem by putting it in perspective with respect to what is already known, including your own previous work on the subject, and explain how you plan to add to current knowledge. Emphasize what is unique about the project. State clearly the theoretic basis for the study if there is one, and indicate briefly how you propose to test the theory.

Your opening paragraph should inspire confidence and convince the reader that what you wish to do warrants careful consideration. Grants are made on a competitive basis, and the merit of your proposed study will be evaluated on the basis of the words you write in your application. If you fail this first test, you may forego the opportunity to explain further, or to defend, any statements which the reader questions or challenges. *Remember that cold print alone must sell your idea.*

MATERIALS AND METHODS

The introduction should lead naturally and smoothly to the section on methods, which is probably the most carefully examined section of the grant application, and therefore one of the most important. If it is superficial and poorly focused, you are not likely to receive approval. Your purpose here is not so much to allow the reader to duplicate your studies, as in a scientific report, but rather to justify the rationale of your study. Describe clearly the method of selection of subjects or materials and the provision of controls for variables that might later be confused with experimental effect. To defend your choice of procedure, you must obviously be familiar with the capabilities and limitations of all available methods. You will want to present alternatives in case your method of choice should prove unsuccessful or inadequate. A discursive presentation of all possible approaches, however, will indicate inadequate critical ability. You should anticipate and compensate for any weaknesses or defects in the method chosen, and should explain precautions to be taken against possible bias. Guard against selecting a sample that is too small to yield reliable results; a biostatistician can be of assistance in determining an adequate size for the sample. Describe also your proposed method of analyzing and interpreting the data, and indicate the anticipated time schedule for the project.

Be sure that your experimental design is rational, ethical, and defensible and that it can yield answers to your stated problem. When the proposed experimentation has ethical implications, it is wise to include a statement of approval of the research protocol from the responsible administrative authority in the parent institution.

Never assume that the reader knows anything of vital significance that you have omitted from your presentation. Even if he knows it, he will not know that *you* know it too unless you tell him. Make certain that you have at least described the sample, the experimental design and procedure, the method of collection and analysis of data, and the time schedule. Considerable time can be lost if the agency must request additional information before a decision can be made.

EXPECTED RESULTS

Following the section on methods is that on expected results, which should be written honestly and objectively. Overstated claims and unwarranted predictions of dramatic "breakthroughs" should be avoided in favor of sober statements based on logic and reason. Try to estimate the potential generality of the results and the basis for your judgment. When tables and graphs summarize data for this section more efficiently than text, they may be used to advantage. Any tests, instruments, publications, films, or other educational material that may be by-products of the project should be noted.

DISCUSSION

In the Discussion, you have an opportunity to show your broad familiarity with the various aspects and implications of the problem and to present the potential significance of the prospective results. Be sure to indicate any distinctive qualities of this research as compared with previous studies. Wild speculation is anathema. This is the place to persuade the reader that your proposed study holds real promise. You will also want to explain your method of assessing the results as a measure of the success of the project. And you may wish to suggest new directions for subsequent or concurrent projects in related research.

REFERENCES

References, instead of being listed at the end, as in a published report, are usually included in the text. Choosing references wisely will testify to your knowledge of the field and your skill in evaluating previous studies. If you omit a reference to previous work because it weakens your argument or shows your study not to be original, the reviewer will assume that you are not adequately informed in your field. A lengthy, uncritical list of references is equally bad. Painstaking accuracy in citing references reflects favorably on the reliability of the applicant.

PERSONNEL

QUALIFICATIONS OF APPLICANT   In addition to the description of the proposed study, the applicant is expected to establish his ability to carry out the study by presenting his qualifications in the form of a curriculum vitae and to provide some evidence of his productivity by attaching a list of his publications.

OTHER PERSONNEL   Applications should, when possible, specify the names, titles, and qualifications of all personnel expected to participate in the project. Indicating the research experience of participants is also helpful. No one should be listed, of course, without his or her explicit permission. Consultants and research teams or agencies that have agreed to provide assistance and cooperation should be included.

FACILITIES

Indicate the space and facilities available for use in this project, including special facilities, such as computers and libraries. Additional space needed should be noted.

BUDGET

The budget should represent a careful estimate of the financial needs to conduct the study. Honesty and common sense are the best guides in preparing the budget. Your ability to direct the study efficiently will be judged partly by your acumen in estimating the costs.

Make certain that you have taken account of fringe benefits for all personnel, such as social security and insurance, and have included an item for overhead, when necessary. Consider whether you will need to pay for transportation of subjects, professional travel for personnel, statistical analyses, preparation or publication of reports, or similar items. Special large items of expense require textual justification. It is wise to have a business manager or purchasing agent review your budget before submission of the proposal.

TABLE OF CONTENTS

A table of contents is welcomed by most readers who wish to review the scope of the proposal before reading it or who, after having completed the first reading, wish to review certain parts of the application. It also serves the writer as another index of the completeness and orderliness of the organization of the proposal, as well as of its strengths and weaknesses. This page obviously cannot be prepared until the application is completed.

REVISION

When you have completed the first draft, lay it aside for a couple of weeks or so—long enough to be able to read it with detachment and objectivity. If you reread it too soon, you may read what you *thought* you wrote, not what you actually put down on paper. Read first for logical order and coherence, accuracy, clarity of purpose, unity of thesis, and consistency; then for grammatical integrity, punctuation, grace of expression, and mellifluence.[3] Note, too, whether the narrative flows smoothly and logically from one statement to the next and from one paragraph to the next. Reading aloud will uncover defective rhythm or cadence, as well as improper balance and emphasis.

If I had to identify the most important literary requisite for grant applications, I would choose precision—precision in choice of words and in their arrangement to convey the intended meaning. Because our language often uses hyperbole and irony, it is tricky and requires considerably more care to convey meaning properly than most of us give it. For example, "fat chance!" actually refers to a slim chance, and "I could care less" means the same as I couldn't care less. Our language also has a certain amount of built-in ambiguity that results from the multiple meanings of certain words and of the different syntactic structures into which they can be placed. For this reason, the writer must be careful about involuntary double meanings, as exemplified in Figure 21-3.

The order of words in English is all-important because position contributes greatly to meaning. There is a vast difference between:

Have you anything left? *and* Have you left anything?

Misplacement of a single phrase can thus play havoc with meaning, as in the following:

The patient would get chest pain *when she lay on her side for over a year*.
Remedied: *For over a year* the patient would get chest pain when she lay on her side.

A dangling construction like this can also distort meaning:

At the age of six, the patient's mother died of cancer.

Obviously, the patient's mother was not six when she died of cancer.

Remedied: When the patient was six years old, her mother died of cancer.

Figure 21-3.   Ambiguity: the group was composed of half men and half women

Close behind precision as an important quality of grant applications is *conciseness*. Not that your composition should be telegraphic or cryptic; rather that you should weed out the pure verbiage, which irritates and bores the reader. You would be surprised how much you can tighten your prose by eliminating the common forms of wordiness. You can, for example, delete introductory deadwood like this without affecting your message in the least:

It is interesting to note that,
It should be pointed out that,
It should be remembered that,
It is significant that,
It is worthy of note that,
It has been demonstrated that,

All these expressions occupy valuable space and time without contributing to meaning. Draw a line through them, and capitalize the next word.

The shortest distance between two points, verbal or spatial, is a straight line, and the more unnecessary words you use to express a thought, the longer it will take your reader to grasp the idea. Look carefully, therefore, for facile but cumbersome circumlocution, and substitute for it the more efficient and more eloquent shorter phrase or word:

Circumlocutory	*Concise*
an excessive amount of	too much
at a high level of productivity	highly productive
at a rapid rate	rapidly
at no time	never
due to the fact that	because
has the capability of	can
in the majority of cases	usually
in close proximity to	near
in view of the fact that	since
it is the opinion of the author	I think
present a picture similar to	resemble
serves the function of being	is

Redundancy slips easily into writing because certain prefabricated expressions are so widely used that they come unbidden to mind. A careful rereading, however, will uncover such superfluity. The people who review your application were selected because of their sharp critical faculties, so you need not use phrases designed for the dimwitted, like:

ancillary aids
basic fundamental essentials (Three times the charm?)
cause was due to, *or* reason was because
cut section
efferent outflow
green in color (As opposed to shape?)
human volunteer (Is there another kind?)
obviate the necessity of
period of time (As opposed to space?)
past memories (As opposed to future?)
psychogenic origin

Tattered phrases, too, impede communication. Exhausted from overuse, these expressions no longer function serviceably, but merely lull the captive reader into somnolence, which may continue throughout the rest of the application. Try, therefore, to root out meaningless banalities like:

a growing body of evidence
a thorough search of the
   literature
a large body of information
   (Figure 21-4)
an in-depth study

appears to be suggestive of the
   possibility that
beyond the scope of this study
by and large
crash program
few and far between

Figure 21-4. Cliché: a large body of information

first and foremost
for all intents and purposes
in the final analysis
last but not least
longitudinal studies

matter of course
new avenues are being explored
suffice it to say
warrants further investigation

Some expressions recur so often as to have become absolutely sterile (see Figures 21-5, 21-6, and 21-7).

Imprecise, capricious language is often a serious impediment to communications.[4] Sprinkling your proposal with unnecessary jargon, as in the following sentence, will not ensure you a position in the inner circle, but rather will expose your lack of resourcefulness:

Effectiveness on the output side of the management prong (information disseminists, documentalists) requires in-depth knowledge of input procedures, relevant sensitivity parameters, and the development of a high index of suspicion of surrogation probabilities.

Vogue words mark the copycat, and whereas they may add spark when first used, their vigor quickly withers from overuse. Moreover, the very evanescence of these terms makes them inappropriate for serious writing.

Figure 21-5. Cliché: the new treatment should prove to be a valuable addition to our therapeutic armamentarium

Remove from your application, therefore, in-words and expressions like: ambience, at this point in time, conceptual framework, dialogue, hopefully, in terms of, input, interface, knowledgeable, meaningful, monolithic, relevant, seminal, simplistic, societal, surrogate, viable, visceral, vis-à-vis.

It is similarly unnecessary to impress the reader with your erudition by using pompous words, esoteric or abstruse phrases, or arcane terms. The discriminating reviewer is not awed or intimidated by such phrases as information data base, media method module, satellite preparation, target population, or learning activity package, nor does your promise of a sequential unit report move him to approbation. The reviewer will, in fact, either scratch his head perplexedly or silently hurl execrations at you if you write such spurious profundity as this:

These perceptual processes are coupled with cognitive functions such as memory and logical thinking which enable adaptive modes of behavior to be deduced from the demands of internal and external reality. Whereas the expressive and defensive functions are sensitive to constancies and may steer the organism toward creating constancy, the adaptive function is oriented toward change.

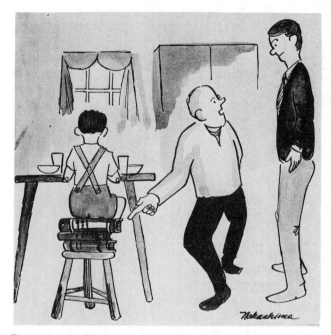

Figure 21-6.   Cliché: this study fills an important gap

Such language seems intended more to conceal than to clarify meaning. The reader will react more favorably if you stick to simple, natural, clear language.

Avoid the freshman's thesaurus-syndrome, in which long synonyms are substituted for simple words in the hope of making writing sound more erudite. This practice has functional hazards, since synonyms usually have at least shades of difference in meaning, and often may have one sense in one context but an entirely different sense in another context. After returning to his native country, a foreign student was writing a letter of appreciation to his American teacher. At the end he wanted to write, "May Heaven preserve you," but was not sure of the meaning of "preserve." First he consulted a dictionary and then a thesaurus and wound up writing, "May Heaven pickle you." Synonyms can trip up even native speakers that way.

Unvarying use of the passive voice, rather than reflecting objectivity, as some suppose, conveys a sense of uncertainty, defensive circumvention, or simple evasion. Rarely does the passive inspire confidence or induce positive action. Try, therefore, to turn around syntactic structures to allow the use of an active, vigorous verb. An accomplice of the passive voice is the abstract

Figure 21-7.   Frozen expression: the monkey proved to be the most suitable model

noun as subject, which, having emasculated the action, requires a weak verb to complete the thought. Note the difference between the following two sentences:

1. *Re-implantation* of the severed arm *was accomplished* by the surgeons.

2. The surgeons *re-implanted the severed arm*.

The visual image of the abstract *re-implantation* in (1) is weak by comparison with that evoked by the active verb *re-implanted*. Releasing the verb in the abstract noun in (1) produces the more forceful and more direct statement in (2).

By examining your proposal carefully with these common literary breaches in mind, you can improve the quality of your writing noticeably. Scrutinize every word, asking yourself if it is necessary to the meaning, if it is the precise word you need to convey your meaning, and if it is in the best position in the sentence to communicate that meaning.[5,6] The more attention you give to these flaws, the more critical you become and the more sensitive to infelicities. The best way, incidentally, to acquire an ear for language is to read good writing—fiction, poetry, drama, prose—whatever you like, but *outside* your field of professional interest.

Figure 21-8. Inept figure of speech: the foot on the other hand appeared swollen

CRITICAL REVIEW BY OTHERS

When you have done the best you can with your proposal, give a clean, double-spaced photocopy to a respected critic who will not withhold constructive criticism for fear of injuring your feelings. Ask him to be ruthless and exacting, for you can be certain that the official reviewers will be just that. Omission of critical information or a flaw in structure may thus be uncovered. A fresh reader can also often point out blind spots that the writer is unaware he has. Inadvertent humor, for example, often escapes the author but is obtrusive to the eye of a new reader:

The foot on the other hand appeared swollen (Figure 21-8).

This disputable statement occurred in a prestigious medical journal:

Autopsied men who ate more than 33 percent of the experimental meals served from entry into the trial to death were more likely to have gallstones (Figure 21-9).

Authors, editors, and proofreaders all overlooked that incongruity.

Finally, let your laboratory director or departmental head review your

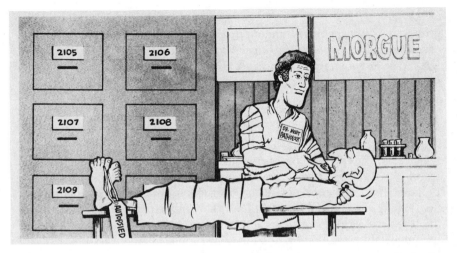

Figure 21-9.   Autopsied men who ate more than 33 percent of the experimental meals

application. Most grants today are made to institutions, not to individuals, and you will therefore probably need his approval and that of your institutional head. It is unfair to expect to rush your application through his and other administrative hands at the final hour; allow enough time for it to be processed in the routine way.

HOW IMPORTANT IS THE ACTUAL WRITING?

Is so much care really necessary in the preparation of an application? After all, you say, is it the applicant's literary proficiency that is under scrutiny or his ability to perform the proposed work? The answer is: both. Some believe, in fact, that the two are inextricably related, and that vague, obscure, disorderly writing reflects the same kind of thinking. Certainly, it seems reasonable to question how clearly a person has conceived his ideas if he is unable to convey them intelligibly. It seems equally reasonable to expect the person who submits a loose, sloppy, ineffectual application to conduct a loose, sloppy, ineffectual study. Moreover, since the utility and value of the results of the study depend on the applicant's ability to present a lucid, coherent report, a poorly written application may signify the writer's inability to communicate any new knowledge which his study uncovers. And the greatest of truths, poorly communicated, remains unconvincing. The history of science and technology is interspersed with valuable observations that were ignored or forgotten for years because the published reports were inscrutable or otherwise ineffectually presented.

APPEARANCE

The physical appearance of the proposal leaves a distinct impression on the reviewer. If the application is messy, illegible, or otherwise unattractive, the reviewer will be irritated or unreceptive even before he begins to read the proposal. Such nonverbal devices as punctuation, spacing, paragraphing, and underlining are valuable signaling tools that will help the reader understand the text more readily. Review the headings, marginal headings, paragraph headings, tables and graphs, and other such elements to make sure they are logical, consistent, and attractively arranged.

CRITERIA FOR EVALUATION OF GRANT PROPOSALS

The peer review system has served science and technology well. Referees usually serve without remuneration, giving generously of their time and effort for the general good. It is a mistake to consider them as adversaries, for their guidance, although not infallible, has generally proved sound in the judicious dispensing of limited funds.

Applications for grants are usually judged on the basis of the following criteria:

The Research Project
Relevance of the research problem to the purpose of the agency
Originality of the concept
Importance of the problem; need for research
Soundness of scientific rationale and approach
Adequacy of experimental design and detail
Feasibility of scope
Appropriateness of budget to proposed work
Suitability of facilities

The Investigator
Education and training
Experience
Research competence

Adequacy of knowledge of research field
Promise
Previous productivity

Before submitting a proposal for consideration, the grant applicant would do well to make sure that these criteria are fulfilled.

SUMMARY AND CONCLUSION

Retrenchment of research funds in recent years has heightened the competition for grants and has emphasized the importance of careful preparation of research proposals. Applying the principles of sound reasoning and clear expression will help the applicant fulfill the criteria of a persuasive proposal: a worthy concept for study, clarity of purpose and scope, and a scientifically valid design for the study, with adequate controls and appropriate method of evaluation—all expressed in orderly, coherent fashion and in simple, concise, readable prose. Thinking through the entire proposal clearly before beginning to write, composing the first draft with attention to logical sequence of ideas rather than grammatical perfection, and removing from the preliminary drafts certain common literary flaws are three steps that will surely enhance the persuasive quality of the proposal.

The applicant should remember that the proposal is a kind of promissory note. It is a mistake to promise mountains and deliver molehills, because the day of reckoning will come—when you will have to write a formal report of your research project. It is through writing, therefore, that both the need for the research and the importance of the results are communicated, and the investigator with literary prowess has a distinct advantage in achieving both of these objectives.

REFERENCES

1. A. Szent-Gyorgyi, Looking Back, *Perspect Biol Med, 15,* pp. 1–5, 1971.

2. L. DeBakey, Releasing Literary Inhibitions in Scientific Reporting, *Can Med Assoc J, 99,* pp. 360–367, 1968.

3. ——, Competent Medical Exposition: The Need and the Attainment, *Surgery, 60,* pp. 1001–1008, 1966.

4. ——, Verbal Eccentricities in Scientific Writing, *N. Engl. J Med, 274,* pp. 437–439, 1966.

5. ——, Every Careless Word That Men Utter: I. The English Language, *Anesth Analg* (Cleve), *49,* pp. 567–574, 1970.

6. ——, Every Careless Word That Men Utter: II. The Language of Science, *Anesth Analg* (Cleve), *49,* pp. 827–832, 1970.

DISCUSSION AND ACTIVITY

1. What does DeBakey identify as the most important literary requisite for grant applications? What obstacles stand in the way of achieving it?

2. Another important requisite is conciseness. What are the sources of wordiness that work against conciseness?

3. What are vogue words and what is wrong with using them?

4. What does DeBakey mean by "the freshman's thesaurus-syndrome"? Why should it be avoided?

5. Why is it helpful to have your writing reviewed by others?

6. Perhaps the most practical and useful piece of advice we can offer you about writing proposals is for you to remember that you must respond to whatever directions and restrictions the proposal evaluator provides. Study the guides for proposals provided by some funding agency (such as the National Science Foundation or the National Institutes of Health) and discuss how prescriptive the proposal submission guidelines are.

7. Write a proposal to do a project for your business and technical writing course (or some other course you are taking) and use the following arrangement for the parts of the proposal:

• statement of problem and aims
• justification of project
• description of method
• expected results
• list of related literature
• qualifications
• timetable for task completion

Selection Twenty-Two

Structure, Content, and Meaning in Technical Manuscripts

HERBERT B. MICHAELSON

Herbert B. Michaelson, in addition to being an editor for IBM Corporation and an associate fellow of the Society for Technical Communication, has taught report writing at Yale and New York University. In this article, Michaelson points out that the quality of any manuscript depends upon its structure, content, and meaning.

The literature on engineering and scientific writing concentrates on three kinds of quality criteria:

1. *Readability and intelligibility:* The favorite target of the communication specialist, who frequently favors word economy and condemns jargon.

2. *Effective expression:* The objective of the English professor, who often stresses sound literary style and clear exposition for maximum communication to readers.

3. *Logical development of technical concepts:* The main approach of those writing experts who have engineering or scientific training.

All discussion and advice on technical writing does not, of course, fall neatly into these three categories. Whatever the viewpoint, however, the usual admonishments on how to write will generally start with the bland assumption that the work to be reported is eminently sound and worth writing about. We will not make such a brave assumption here because, as is well known, the glitter of today's technical literature is not all pure gold. Indeed, the problem is not how to swell further the flood of engineering

Technical Communication XV (Second Quarter 1968), 15–18. Reprinted by permission of The Society for Technical Communication, 1010 Vermont Avenue, N.W., Washington, D.C. 20005.

reports (which no one can keep up with anyhow), but rather how to direct the flow of the better papers into useful channels. Many technical manuscripts are too lengthy for their purpose, and the labor of the writing and heavy rewriting is not always justified by the quality of the work reported. On the other hand, many fine potential contributors to the literature are never written—not because the work is not worth publishing, but because the author's skills are undeveloped. For him the job of composing an acceptable report, getting it referred and published is too time-consuming and demanding to be worthwhile.

When we consider that tremendous effort is expended on writing about routine and trivial topics and that some papers on excellent work never materialize, we must reconsider the usual advice of experts on technical writing. In this article we will try a new approach by examining three properties of a manuscript:

1. *Structure:* The way a technical report is put together to show the relative importance of various aspects of the author's work.

2. *Content:* The exact nature and purpose of the work reported and the author's interpretations.

3. *Meaning:* The significance of the manuscript to a given class of readers, i.e., the efficiency of the author's communication of technical ideas.

These three properties are by no means separate and independent. The way they interact and the effect on the quality of a technical manuscript are the subjects of our discussion.

STRUCTURE

By the "structure" of a manuscript we mean the way it is put together—its division into sections and the disposition of its tables and illustrations. The structure, in its truest sense, is the foundation of a manuscript and the author's choice of a structure determines at the start the whole character of the paper.

The first structural detail is the beginning section of the manuscript. Without a sound framework the manuscript will mislead the reader in several ways. If it lacks a good orientation of the problem in the early sections, the average reader is at a disadvantage to understand the real purpose of what follows. When one section of a paper does not logically follow another, or tabular material is badly designed, or charts and photographs are poorly chosen, the popular remedies for poor technical writing don't help the reader. Short words and short sentences do not improve the situation. Avoiding jargon accomplishes nothing in this case. Precise grammar, spelling, and punctuation are of no avail. The use of well-turned phrases and a style of language meticulously chosen to suit the level of the readers will not solve the inherent problems here. The precision and style of language, of course, are important but are secondary to a careful overall shaping of information and the exposure of technical concepts in appropriate portions of the manuscript.

A sound structure tends to show logic and merit in the work being reported. The structure is sometimes revealed in the abstract. If the abstract comes directly to the point, showing the purpose and content of the paper, the writer is off to a good start. On the other hand, a confused, vague, or wordy abstract will frequently be followed by a poorly constructed paper.

An ideal type of brief, informative abstract, adaptable to any kind of engineering or scientific manuscript, may consist of only three sentences: (1) a pithy statement of the problem, (2) an identification of the author's approach to the problem, and (3) a short statement of the essential results. The structure of such an abstract is similar to the classic building block of a manuscript: Introduction-Body-Conclusion.

As for the manuscript itself, the "fine structure" depends a great deal on the topic and on the purpose of the work. A few specimen outlines are shown in Figure 22-1. There are, of course, many other varieties, and the choice of an appropriate structure depends on the author's sense of order and balance. These forms of technical manuscripts appear stereotyped only to the novice. To the experienced author an imaginative approach to a basic outline, proportioned to emphasize the author's more important contributions, is the basis of an effectively written manuscript.

A sound structure, however, does far more than provide proper order and emphasis in a paper. In subtle ways, *choosing a structure actually affects the content of the manuscript*. For example, the standard type of Introduction in a piece of technical writing demands a statement of the problem; an explanation of what has been previously done in this field, and briefly who has published the important developments; some remarks about how the present approach differs from what others have done; and perhaps a brief advance view of what will be discussed in the manuscript. This scheme of writing the introductory portion is suggested in Figure 22-2.

Following such a conventional routine will require the author to assemble several classes of information for his Introduction. He may well have to go back to the library for more information, or he might have to check with his colleagues to ensure that he is reporting the correct background information. In this sense, the *content* of the introductory paragraphs can be affected by a set of conventional structural requirements. Of course, we are not suggesting a stereotyped Introduction that answers the items in Figure 22-2 in monotonous succession. These general types of information, however, tend to find their way into most manuscripts.

Another effect of structure on actual content sometimes occurs when an

Abstract
Introduction
Theory
Analysis
Experimental Results
Conclusions
References
Appendix

Summary
Statement of Problem
Design Approach
Derivation of Design
   Equations
Applications
Illustrative Examples
Conclusions
Bibliography

Abstract
Contract Requirements
Historical Background
Design Principles
Development of
   Prototype Model
Properties
Life Tests
Reliability
Conclusions
Appendix

Figure 22-1.   Representative outlines of technical manuscripts

author is attempting to arrange data in some logical way. Redesigning a table or chart for clarity can unexpectedly reveal new trends and inspire new interpretations; and rewriting for better literary structure involves rethinking. All of us at one time or another have gained a new insight into the nature of a technical concept by having to explain it in writing.

By the same virtue, outlining a manuscript and building it according to a plan will sometimes reveal an unintended gap in the work. Many an author,

Abstract
A. The purpose of the paper
B. A very brief summary of the results
C. General statement of the significance of the work and its applications

I. **Introduction**
   A. A statement of the exact nature of the problem
   B. The background of previous work on this problem, including published work
   C. The purpose of this paper
   D. The method by which the problem will be attacked
   E. The significance of this work; explanation of its **novel** features
   F. A statement of the organization of the material in the paper

II. **Body of the Paper**
    (The organization of this main part of the paper is left to the discretion of the author. The information should be presented in some logical sequence, the major points should be emphasized with suitable illustrations, and the less important ideas subordinated in some appropriate way. This portion of the paper should be styled for the specialist and should not be "watered down" for the general reader.)

III. **Conclusions**
     A. A statement of how the original objectives were met
     B. Summary and evaluation of the work done
     C. The limitations of this work
     D. Advantages over previous work in this field
     E. Applications and general significance of the results

Figure 22-2. Suggested outline for journal's papers

while writing a manuscript, has suddenly sensed a missing piece of information and has rushed back to the laboratory for more data! And many an author has acquired a new understanding of what he had been doing in the laboratory when he struggled to shape up an evaluating Conclusion for his report. In these subtle ways, then, the attempts to structure a paper can actually contribute to its content.

Structure has another obvious but important property—it also affects the over-all significance to the reader, i.e., the meaning. Since any scientist or engineer finds large quantities of literature in his field of interest, he appraises the over-all meaning of any paper by scanning it before he actually reads it. In scanning, he observes its structure, i.e., he looks at the headings, senses the relative length of the various sections, glances at the illustrations, and forms an opinion of what is in the paper and what is significant about it. The way the blocks of information are fitted together affects the over-all meaning. The sequence of sections, for example, shows the author's line of thinking and indicates how his concepts or descriptions will develop. The relative length of each section gives an impression of its relative importance, i.e., lengthy sections imply a heavy percentage of the author's effort and may lead the reader to attach undue importance to those portions. In this sense, the structure of the manuscript actually colors the meaning that gets through to the reader.

We can see, then, why the structure of a report is a terribly basic consideration—one that unfortunately may be cast aside by the author who claims that he doesn't need a written outline or that he can't work with one because it continues to change as the paper develops. The fact is, especially for those who have difficulty in writing, that an outline carefully thought out will establish the desired structure and become the key to the character of the manuscript. In the overall structure, perhaps far more than in the author's skill with language, is evidence of an author's judgment and his ability to highlight the significant aspects of his work and to deemphasize the less important ones.

CONTENT

Our definition of the content of a paper was "the exact nature and real purpose of the work reported." We use the adjective *exact* because the real nature and intent of any given manuscript are usually subject to interpretation. Clarity in technical writing can be a surprisingly elusive quality—what is clear to one class of readers might not be at all clear to another. For this reason the first requisite of a good engineering or scientific paper is an adequate statement of the problem at hand and a lucid explanation of the background of the work. Without this kind of preparation for what is to come in the main body of the report, the average reader might have to peel off layer after layer of detailed exposition before he gets to the core of the author's real contribution. But in technical writing, unlike other forms of literature, it is considered bad form not to come immediately to the point.

Somewhere early in the manuscript, then, the precise nature of the work should be identified. In the whole realm of scientific and engineering literature there are really only a few kinds of content. These might be classed in general as follows: the original contribution (either theoretical or experimental); the application or analysis of known concepts; the description of a device, system, or methods; the properties of a material or process; a report of progress or completion of a project; and the tutorial review.

Confusions as to actual content arise in several ways. If an application paper, for example, is written in the flavor of an original contribution, with the vague implication that the information is being published for the first time, the reader can be misled. A writer seldom intends to be misleading but can neglect to place the proper labels on his work.

Another type of confusion arises in whether the work reported is complete or fragmentary. An example is a paper describing a device, which appears at first glance to be a completed and practical design but on closer examination turns out to be merely a progress report, with many gaps and unfinished areas of development.

Technical papers should give some clear indication of the status of the results. If the work can be called complete (and few papers are really "complete"), it is a well rounded investigation, including a discussion of all the pertinent aspects, with analyses and proven results. If the work is fragmentary, it has gaps, inadequacies, or tentative results, or is merely one phase of a segmented work project. Such manuscripts need a section devoted to the limitations of the results. Any technical paper that lacks this self-imposed critique implies that it has adequately covered everything there is to say about the subject and, therefore, that the study is complete.

The real purpose of a manuscript may differ from its *apparent* purpose. For example, the actual purpose might be the survival of the author in an academic world of publish-or-perish! Or it might be to justify the expenditure of project funds. Or at the other end of the spectrum of motivation the purpose might be to expose a new concept to the world of science. Since the actual purpose of the work is included in our definition of content, we might list here a variety of the true objectives of technical reports and papers:

1. *To reveal* a new concept or application for the edification of the scientific or engineering community.

2. *To evaluate* work that has been previously reported: to compare, to analyze, to determine relative importance, and to investigate feasibility of ideas or processes.

3. *To recommend* a solution to a given problem; usually involves an assembly of pertinent technical information and an analysis of unresolved questions.

4. *To inform* on the progress of project work and how the objectives have been approached.

5. *To instruct* in technical principles, showing the extent of existing knowledge and methods.

6. *To support marketing* by providing information about a product or service.

7. *To justify work or funds* on a specific project, or to request new funds or other support.

8. *To get on record* so as to establish a reputation for competence, or merely to obtain credit for work done.

These eight reasons for preparing a technical report are not entirely distinct from one another, and manuscripts are often written for several such reasons. The manuscript will usually show evidence of how one of these reasons predominates the author's thinking. For example, a paper on the engineering design of a new system might have a long section explaining the difficulty of the original problem if the author's main purpose is to bolster his reputation for competence. Or the report might have a lengthy, detailed section dealing with the background of related work if his chief purpose is tutorial. Or it might contain an extensive analytical section if his main purpose is to evaluate the merit of the design. Indeed, an author's real motives and the main emphasis in the paper can sometimes be judged better by observing the relative length of its sections than by reading the abstract!

We pointed out previously that the structure adopted by the author has certain subtle effects on the content of a paper. The intended content, on the other hand, also influences a writer's choice of structure but in a way that is anything but subtle. Some of these effects become rather obvious when one considers that a heavily theoretical paper will have one or more sections showing how the ideas are used in practice. And a report that seeks the solution of an engineering problem will generally have a Recommendations section, and so on. Moreover, each engineering or scientific discipline tends to demand its own structural form, and so attempts to prescribe a standard outline for all reports and papers are usually futile. Content dictates structure, which then develops according to the taste and judgment of the author.

The content of a paper also has a rather drastic effect on what we have called "meaning," i.e., the significance to a given class of readers. Here is a crucial point that is sometimes underestimated even by the most experienced technical authors. Unless the writer can get a clear idea beforehand of his audience, he is not in a favorable position to decide on the main content of his paper. What is technically meaningful to the specialist may not always be meaningful to management who must base decisions on a technical report. Another aspect of technical content concerns today's overlapping fields in science and engineering. For example, when a report will be read by a physicist, a chemist, and a device engineer who are working on the same project, the technical content can be more meaningful to one class of specialists than another. In the long run, the judicious author finds himself making a compromise in his choice of content to suit his various readers.

MEANING

The most controversial and troublesome aspect of technical literature is not the structure or content of manuscripts; rather, it is how efficiently a manuscript communicates information to readers—a quality that we have called "meaning." The most frustrating single problem for the technical reader is ambiguity. There are three kinds: *linguistic ambiguity*, which concerns multiple word meanings and connotations; *logical ambiguity*, consisting of gaps or contradictions in technical concepts; and *structural ambiguity*, an uncertainty of over-all significance due to poor organization of the manuscript. These faults, of course, are the favorite targets of editors and referees; any manuscript free of such ambiguities has a great deal in its favor. As we have already seen, however, structure and meaning alone are no assurance of quality—that third essential element of *content* is as important as the first two. Indeed, a paper can be structured in fine style and written in the clearest language and with sound logic—but if the nature of the work reported is trivial, the manuscript is trivial!

We have already suggested how structure and content can affect the meaning of a manuscript. There are reciprocal effects, too. Most technical work, particularly research and development, involves tackling a problem and finding a solution. Sometimes the problem is relatively simple but demands involved or elegant methods for solving it. Or the problem may be terribly complex but can be solved in a dramatically simple way. Whatever the situation, the evidence of technical skill eventually appears in a manuscript of some kind—either in laboratory notes or in formal writing, such as a report of a paper for publication.

In these circumstances, the real "meaning" of an author's findings is frequently in a fluid state. Laboratory work or analytical studies can be routine for long periods of time, but results may come in fits and starts. New directions and new thinking come unexpectedly in any technical project—sometimes even after the project is supposedly finished and while a report is being written. The alert and imaginative author finds that "meaning" grows with a manuscript and that refining his ideas involves reshaping his writing. The author may not be fully aware that he is actually reinterpreting his months of work in the laboratory. But, aware of it or not, the author finds new shades of meaning in his work as he writes and rewrites. In this way, meaning affects the structure and content of his paper.

CONCLUSION

We have attempted to show that the over-all quality of a piece of technical writing does not depend on such simple criteria as its literary style or the technical merit of the author's work. Even if a paper is "clearly" written and polished in the best expository style, it will not be considered a high-quality paper if its content is trivial or outmoded. On the other hand, an excellent piece of technical work might be buried in an avalanche of poor, confused writing. But even if a paper has good content and literary style, any attempts to define its quality are further complicated by the different viewpoints and needs of various sets of readers.

Our approach to this over-all question was to suggest three benchmarks of quality for any technical manuscript: structure, content, and meaning. This combination of three criteria accounts for the questions we have just posed on expository style, merit of the work, and the needs of various readers.

The interactions of the three criteria are suggested in Figure 22-3. Since structure, content, and meaning depend on each other in several important ways, a weakness in any of the three will certainly affect the other two. For those who appraise manuscripts, the diagram might offer a new and useful way of looking at a paper and judging its ultimate worth. The diagram can have another purpose, however: It can illustrate for authors and editors the real-life process of technical writing. Although the time-honored recommendations of "clear" writing, sound literary style, and logical development of technical concepts have their special value, Figure 22-3 offers a broader view of the writing process.

The first question that occurs in any interpretation of Figure 22-3 is: Where does the writer enter the situation and where does he leave it? This would depend on the individual, but the likely place to enter the circle of interaction would be at Content, i.e., a consideration of the nature of the work to be reported and the type of manuscript to be written—a short, informal note, a formal report, or a full paper for journal publication.

The next step would be a careful shaping up of the Structure, or outline, according to the main purposes of the manuscript. The last consideration should probably be Meaning—the choice of proper emphasis and flavor for the manuscript, according to the class of readers who will primarily be concerned with it.

The only practical way of testing the rough draft for Meaning is to send it for review to some competent person in the intended class of readers. Later the author can revise the draft on the basis of at least one such evaluation.

Our viewpoint here has been that of writing a high-quality manuscript in any of the disciplines of science or engineering, and the writing process summarized in Figure 22-3 is our special view of how a good paper matures.

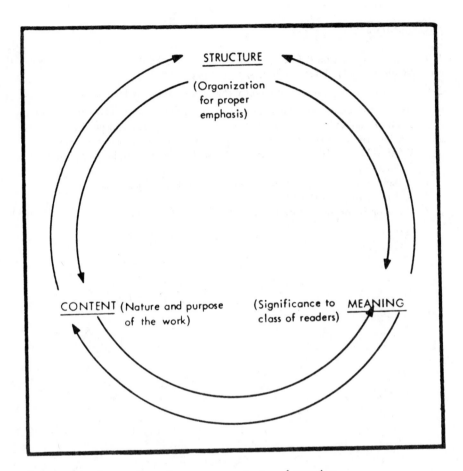

Figure 22-3.   Interactions of structure, content, and meaning

Needless to say it does not apply to the manuscript that is mechanically constructed by a plodding, unimaginative author who feels that he has finished his work in the laboratory and proceeds to "write it up for publication," tossing off the first draft and then abandoning further effort or interest in his work. The readers of such a manuscript inevitably sense the routine flavor and the lack of inspiration or excitement, even though the manuscript may boast of technical accomplishment. But the paper of quality, written in the heat of the interactions shown in the figure, will have the earmarks of fresh insights and of the inspired thinking that characterizes good work.

DISCUSSION AND ACTIVITY

1. How does Michaelson define "structure"? How does he define "content"? What are the eight possible objectives of reports and papers?

2. What are the three kinds of ambiguity that cause problems for readers?

3. According to Michaelson, the quality of any manuscript depends on its structure, content, and meaning. Read an article from a business or technical periodical and analyze it in terms of these three properties, answering the following questions:

    a. What techniques does the writer use to clue the reader in on the overall shaping of the material?

    b. Which of the eight possible objectives of reports and papers are apparently behind the writing of the article?

    c. What problems of ambiguity are present?

Selection Twenty-Three

How to Organize a Research Report for Management

RAYMOND A. ROGERS

Raymond A. Rogers is a senior research chemist at Union Oil Company's Brea Research Center, Brea, California, where he devotes full time to editing research reports and counseling research writers. Rogers says that the busy manager is likely to be impatient with the traditional scheme of reporting. The solution is to organize a report along the lines of a newspaper story, starting with the most important points.

Let's say that you have just finished a research project and are getting ready to write the first draft of a report to the manager of your group. For several weeks, starting even before your work was completed, you have been speculating about what you want to tell your manager and the other management men who will see this report. Several days ago, when your ideas began to take firm shape, you made a tentative outline to help you visualize those ideas and to find out if they would hang together. You have been juggling the topics of the outline, changing some of them, filling in gaps between them, trying them out in different sequences, and you seem to have found an arrangement that brings out a central pattern of meaning among them. You therefore think you have figured out the significance of your results, and you know roughly what you want to write, but you haven't yet decided how to go about it.

How, for example, are you going to start telling the story? What is your opening statement going to be?

Technical Communication 20 (First Quarter 1973), 8–9. Reprinted by permission of The Society for Technical Communication, 1010 Vermont Avenue, N.W., Washington, D.C. 20005.

START WITH THE MOST IMPORTANT

At this point, the telephone rings. Your immediate supervisor, who is on a business trip, is calling from a distant city. He says he is on his way to a meeting where he, along with other company executives, will make a certain critical decision. The meeting starts in 10 minutes. The results of your research have a bearing on the decision. Of course, your supervisor knows all about the problem you are working on and the objective of the work. Before he started his trip, he had known that you were getting along toward the end of your laboratory investigation, and he hopes you can tell him something that will help him in the meeting. You have only 3 or 4 minutes to talk.

What are you going to say?

Obviously, you're going to tell him whatever you consider most important and significant about your findings. And, since you have only a few minutes to talk, you're not going to waste time building him up for your main message. You're just going to give it to him—straight.

As soon as you finish talking, you can hang up the receiver and start writing your report. What you have just said to your boss should appear at the beginning of this report, immediately after brief statements of the problem that led to the work and your objective in doing it. These preliminary statements have to be made before you can really start reporting. They may not be needed, of course, by your own supervisor, but they will undoubtedly be needed by the other managers who will see your report.

YOUR BOSS IS FACING AN EMERGENCY

Let's do a little more supposing. Let's say you had finished your report and had turned it over to your supervisor before he started on his trip. He might have read it on the plane but he didn't. He had many other things to work on while he was traveling, and some of these other things seemed more immediately important than your report. He didn't know, until 10 minutes before the big meeting, that the decision involving your work was going to come up. As soon as he found this out, he snatched your report out of his briefcase and did the best he could with it in the time he had.

If you wrote that report as some reports are written, starting with a long historical-theoretical-general introduction, and then continuing with a detailed discussion that gradually built up to your big message, your boss was in trouble. But if you started out with a straightforward, unadorned statement of your main message, he was able to go into the meeting well prepared.

When you are writing a report to management, it always makes sense to present your main message at the beginning. You have to assume that your report is going to be read during an emergency of some kind. Most executives seem to operate in a sort of routinely continuing, sustained emergency that varies from acute crisis to the normal, everyday administrative quandary. They are facing either a critical challenge like the one we have been describing or the chronic, low-grade predicament of not ever having enough time to do all the things they need to do.

It is obvious that our most-important-thing-first approach is a big help to a man facing an acute crisis. When you stop to reflect, it is *also* obvious that this straightforward approach is a big help under any circumstances to a man who is chronically short of time: it means that he won't have to work so hard or so long to understand what you have to tell him.

START WITH OUTCOME OF WORK

When we write about a conventional research project that involves laboratory or pilot-plant work, the findings that an administrator can apply will be either our results, some conclusions that are based on results, or some recommendations that are based on conclusions and results. These are the usual outcomes of such a project.

Some of our reports, however, have nothing to do with conventional projects. When you write a different kind of report, you will have a different kind of outcome to present at the beginning.

If, for example, you have been studying an established process in order to evaluate it, present your evaluation first. If you are writing to offer a proposal or suggestion, start out with the proposal or suggestion. If you are offering your thoughts about some controversial question, you might want to call your opening statement a viewpoint, an opinion, an interpretation, a thesis, a theme, or even an argument. If you are reporting some other kind of investigation or speculation that had some other kind of outcome, begin with that outcome, whatever you choose to call it.

The most important thing you have to say comes first, the second most important thing comes second, the third most important thing comes third, and so on. The remaining topics in your outline and report are presented in the order of their decreasing importance.

When you put a report together in this way, you are making it possible for a reader *who doesn't have time to finish your whole report* to get maximum information by reading straight ahead. You are also increasing the possibility that he will be able to find, without much searching, somewhere on the first two or three pages, any special item of information that he needs.

If you are reporting a conventional research project involving a laboratory investigation, you might use a sequence something like the one shown in Figure 23-1. For this kind of report, as we see immediately, the sequence used in reporting turns out to be roughly the opposite of the one that is followed when the work is done. However, it is not exactly the opposite. Your work started with a problem and an objective, and your statement of

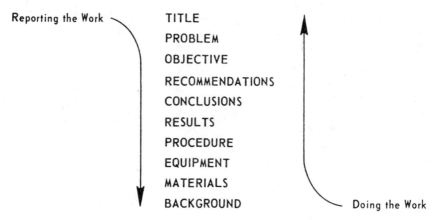

Figure 23-1. Sequence used in reporting versus chronology of the actual work

these items must come first in the report. Likewise, your discussion of the other items will not always be exactly the reverse of your work sequence. Simply reversing this sequence is a good way to start organizing your outline. But often some adjustments of the reverse-work sequence will have to be made to get the items exactly in the order of their decreasing importance.

GIVE THE DETAILS TOO

Let's take another quick look at the meeting in that distant city where your supervisor was using, as an emergency source of information, the report that you turned over to him before he started his trip. He was able to extract your main message quickly from your report and pass it along to the other administrators, but this turned out not to be quite enough to satisfy them. Several members of the group started asking questions about the details of your work. Executives can sometimes be very nasty about inquiring into details. Because your boss had not yet read your report, he found it necessary to search out the requested information in a hurry. Again, if you used the reporting method we just described, he was able to do this without any trouble at all.

A glance at the reporting sequence given in Figure 23-1 shows that you can easily include in this kind of report as much detail as anyone might need. There are plenty of places for it. In fact, you can become as scholarly and expansive as you wish—after you get past the critical beginning parts. In the later sections, you can work in all the particulars you consider appropriate, even enough to satisfy the requirements of other researchers and students, whose interests are quite different from the interests of managers.

Instead of	We might write
PROBLEM	**PLUGGING LOWERS RECOVERY BY 15 PERCENT**
EQUIPMENT	**DISPLACEMENT-TEST APPARATUS**
CONCLUSION	**INTERFACIAL TENSION BETWEEN BRINE AND RATHBONE CRUDE OIL IS pH-DEPENDENT**
MATERIALS	**TYPES OF CORES USED IN THE TESTS**
PROCEDURE	**CHARACTERIZATION OF CARBOXYLIC ACIDS AS HYDROCARBONS**
RESULTS	**SODIUM HYDROXIDE ADSORPTION ON NEVADA-130 SAND**

Figure 23-2. Informative headings

Not only is there room for a lot of information, but particular items are easy to locate in a report that has been put together in this way. With only a little further reflection, you can make such items even easier to find.

USE MANY HEADINGS

A most effective way of making information easily retrievable is to use lots of headings and subheadings: to construct a kind of index of your report, and embody this index in the report itself. Subheadings, in particular, can help you keep detail straightened out and easy to locate.

USE INFORMATIVE HEADINGS

There is nothing wrong with the label headings shown in Figure 23-1, but, to a harried executive, informational headings can be far more helpful. The randomly selected examples given in Figure 23-2 show that phrases and short sentences can function very effectively as headings. Phrases and sentences used as headings may not be necessary, appropriate, or effective for all parts of your report, but there is nothing wrong with using some of them right along with some label headings.

USE AN INFORMATIVE TITLE

Sometimes it is possible to pack a surprising amount of information into a title without making it too long. You can usually formulate a good title with nouns and phrases, but it is far better to work in at least one verb, converting

Instead of	We might write
SELECTIVE PLUGGING OF STRATIFIED SYSTEMS	SELECTIVE PLUGGING CAN INCREASE OIL RECOVERY FROM STRATIFIED SYSTEMS
EVALUATION OF WASTE SULFITE LIQUORS	PAPER-MILL WASTE SULFITE LIQUORS DID NOT PREVENT CLAY-SWELLING DAMAGE TO WATER-SENSITIVE CORES
PUMPABILITY OF RED-FEATHER CRUDE OIL	CARBON DIOXIDE INCREASES THE PUMPABILITY OF REDFEATHER CRUDE OIL
SOLUSEAL AS A FLUID-LOSS AGENT	SOLUSEAL IS A GOOD FLUID-LOSS AGENT, BUT IT MAY CAUSE EMULSION PROBLEMS
FRACTURING WITH A GELLED DISTILLATE	PARAFFIN DEPOSITION WOULD NOT BE CAUSED BY FRACTURING WITH A GELLED DISTILLATE FROM A GAS CONDENSATE FIELD

Figure 23-3. Informative titles

the title into a sentence. Examples are shown in Figure 23-3. An informative title helps an administrator get oriented quickly to what he is going to read, and it helps other potential readers decide whether or not they should study your report. Sentence-titles are especially helpful to people who are searching for information in the library.

YOU PROBABLY DON'T NEED A SEPARATE INTRODUCTION

Usually your statements of problem and objective are all the introduction you will need. For some reports, however, it is possible that management will need some additional introductory information in order to understand or appreciate fully the "most important thing" you have to say. Nevertheless, even in this case, don't write a separate introductory section—unless you can keep it *very* short. Try to blend your introductory remarks right in with your statement of the "most important thing."

The trouble with a separate introduction is that it tends to defeat its own purpose by becoming swollen with historical, theoretical, and general information that is not truly introductory. Instead of helping your reader get into the report, it can easily become a barrier to him.

If you wish to include general information, call it Background and place it at the end of the report. There it won't get in anybody's way, but it can easily be found if it is needed.

YOU DON'T NEED A SUMMARY

Even though you have placed your main message at the beginning of the report, you may still feel, out of habit, that you should start with a summary.

At first, some of the management men who read your reports will undoubt-edly feel this way too. Administrators have become accustomed to relying on summaries. Every executive wants to believe that he can be "briefed" in a couple of minutes, no matter how complex the message he is receiving, and report writers feel that they should try to meet this expectation.

Thus the big trouble with the summary is that administrators depend too heavily upon it. The word *summary* encourages them to assume that they can encompass and comprehend the whole report by merely reading one short section at the beginning. They tend to treat the summary as a thumbnail, nutshell, detachable version of the report—*as a substitute for the whole report*. Executives too often take it for granted that they can read the summary and skip the report.

This is, for the most part, too much to expect from summaries and also from the research people who write reports. Nutshell, detachable substitutes for reports may possibly sometimes be produced by exceptional writers, but not by ordinary ones like you and me. It is rare indeed to find a summary that can take the place of the whole report. Since this is the case, the tendency of administrators to depend so heavily on summaries is dangerous. The execu-tive who considers himself adequately "briefed" because he has read a sum-mary may be ignorant of the information that he needs most.

Rather than attempt to write a detachable summary, we present our most important findings under the most specific and descriptive headings we can devise. Although these main findings are placed at the beginning, where the summary would be found, our headings do not encourage administrators to limit their attention to the main findings alone. The executive who reads brief sections with such headings as Problem, Conclusions, Recommendations—or other more descriptive headings—is not likely to become bemused into thinking that these brief sections can be trusted to represent the whole report. On the contrary, these beginning sections are likely to whet his curiosity and interest and lead him to look at other parts of the report that are of interest to him.

Since we have used a great many specific and descriptive headings throughout the report, we have made it easy for any administrator to find his way around in the *whole* report, to locate and absorb quickly the information—including the detailed information—that he needs most, to skip over any parts that he does not need, and to check back and review important points after his first perusal. We have, in effect, made the *whole report* as easily comprehensible as a summary.

Some people say they don't expect a summary to substitute for the whole report—just for the most important parts of it. They argue that we should retain summaries and include in them only conclusions, recommendations, and objectives. But why lump together the most important statements of a report under a nondescriptive heading like Summary? Let's separate these most important parts and call them by their right names so that they can be easily recognized and differentiated.

The sequence we are recommending for the beginning of a report has none of the disadvantages of a summary and all of the advantages—and in addition it has flexibility. Different executives need different amounts and kinds of emergency information, and our sequence allows any administrator to decide for himself when he has read enough to become sufficiently "briefed."

DISCUSSION AND ACTIVITY

1. Rogers recommends that report writers start by presenting their most important results, information, and findings, and that they arrange them in the order of decreasing importance. Read several reports or articles in your major area of study and determine where writers place their most important results, information, and findings, and how they have arranged them.

2. Rogers recommends that report writers should expand headings and subheadings into informative statements (for instance, a heading such as "Advantages of Asphalt" would become "Asphalt Attractive in Cost, Installation, and Maintenance"). Find a report or article that contains headings and subheadings that could be made more informative. Place the original headings and subheadings in a column on the left side of a sheet of paper and your expanded headings and subheadings in a column to the right.

Selection Twenty-Four

The Corporation Reports to the Reader

A. M. I. FISKIN

A. M. I. Fiskin has taught report writing at Pennsylvania State University and served as a consultant to industry, including the Oak Ridge National Laboratory. Pointing out that stockholders represent many levels of economic sophistication, Fiskin finds that corporate annual reports are written to increase the understanding and pride of ownership of stockholders. Each section of a corporate annual report must be written for a specific segment of stockholders.

The corporation annual report is a strange document, colorful and decorative, although sometimes, unfortunately, tasteless. Very little of what it contains is required by law, and the financial sections and Auditors Opinions that are required do not provide the incisive information that appears in a prospectus. Although simplified in presentation, it is often left unread because it is still too complicated. Despite these limitations, however, everyone feels that the Report is a good thing and that it should be widely distributed.

For whom are Annual Reports written? The answer is not entirely certain. Of course Reports go to stockholders; they are also often sent to bondholders. Sometimes, but not always, they are distributed to employees; they are frequently used in recruiting new employees; on occasion they are piled high in brokers' offices; and each year a group of almost 200 companies advertises together in appropriate newspapers, enticing readers for their Reports. Out of confused purposes there has developed a format, a literary genre we might say, peculiar to our times. It may well be that this format should be dis-

Journal of Business Communication XI (Winter 1974), 43–50. Reprinted by permission of the publisher.

carded, that another device for disseminating the necessary information might be more attractive, more useful, and sometimes more honest. But the Corporation Annual Report is here and therefore requires, even if sometimes it does not merit, examination.

The creators of these documents are not always satisfied with their own work and may judge themselves too harshly. The fact is that they are composing intrinsically difficult documents that have to reach too many different levels. Within each of the groups mentioned are readers of varying sophistication, and, in fact, the differences are probably most clearly marked within the primary audience, the stockholders themselves. Among them are knowledgeable investors and bemused speculators, as well as economic naifs who have inherited a few or many shares and think stock certificates are dangerous scrolls. Clearly no Report can ever succeed to the point that every stockholder reads it and is properly edified. Although Reports have improved over the years, it seems that the efforts to improve them have no direction, and good results are achieved almost by accident. In this paper I hope to suggest a way of defining the problem in terms that all professional communicators accept.

Every textbook on business writing tells us to be clearly conscious of what it is we want to communicate. Every textbook, quite properly, tells us to be conscious of our reader. When we begin to face the actual problems of composition, we all learn that there are sometimes different problems of composition, we all learn that there are sometimes different readers, with different sets of needs. We then begin to think of double reports or the virtues of appendices. The Company Annual Report does not have two kinds of readers; it has many kinds of readers. There is no way the writer can reasonably expect to reach all of them in all parts of the document. But all the readers have two needs. They want information about the company's activities, both the good news and the bad and they want to feel pride in their company. Management, incidentally, is equally anxious to create this feeling. Success for the writer consists in making each section a little more meaningful for a slightly greater number of those who are that section's natural audience.

THE SECTIONS OF THE REPORT

It is an unfortunate fact that many stockholders never open their Reports. Those who do probably find that the Report includes these sections, usually arranged in this order, although the titles may be different.

- Highlights
- President's Letter
- Operations
- Financial Statements (Usually followed by Notes to the Statement)
- Management

HIGHLIGHTS

As "Highlights" usually appear on the first page or the inside cover, the recipient reads this section if he reads anything. The "Highlights" section explains the company's finances at the lowest level of oversimplification, and usually compares the activities of the current year with those of the year before. Occasionally "Highlights" list as few as five items: Sales, Income before Taxes, Income after Taxes, Earnings per Share and Stockholders Equity. A table may be laid out attractively, boxed in one quadrant of the page, and surrounded by acres of white or colored space. Even truncated as it is, this section may indeed be all that some readers want: they see whether the company has made more money or less than the year before; they register the appropriate emotion, and they are satisfied.

Often the writers miss an opportunity to increase their effectiveness. Because so many read this section and no other, the writer should include as many items as he thinks the reader would accept before quailing at the mysteries of high finance. The reader would surely accept eight, ten, or twelve pieces of information. And each would add to his understanding of the company.

Occasionally Reports include at this point, or elsewhere, summary information about important activities of the company: development of a new product, opening of new markets, completion of a merger, or other items that are truly "highlights." These events should be presented in narrative, ranging in length from a short sentence to a short paragraph. They catch the attention of even the most timid reader and help him to understand and perhaps identify more closely with the corporation.

THE PRESIDENT'S LETTER

It is perhaps unfortunate that the "President's Letter" comes as early as it does in the report, right after "Highlights." Far too many readers look at the first two paragraphs and then put the report aside, never to return to it. It is fitting, of course, that the head of the company address the stockholders, but the letter is usually written in a style so graceless that we are forced to believe that the president actually wrote it himself! Given the facts of corporate life, it then suffers immunity from criticism, correction, or even comment. Frequently the "Letter" only repeats awkwardly the information already presented in columns in the facing-page "Highlights." The huddled figures of millions of dollars of sales, percentages of increase and ratios result in a passage of awkward unreadable prose.

In a routine year the "Letter" is simply a desirable ritual. When there are new things to report, it can be interesting and convincing; information in it is, after all, signed and therefore official. It should, however, discuss the new developments briefly. Details are properly developed in other sections of the Report.

OPERATIONS

The "Operations" section, as it is usually called, is the heart of the Annual Report. It is not a legal requirement, nor is it read by all recipients, but it is read by the most interested of the shareholders. It is there, in fact, for the ideal reader, and writing for him should be a source of satisfaction for those who compose Reports. He is naturally interested in profits, but he is also interested in the products the company makes, in the services it performs, in employee relationships, in the contribution of the company to society and to mankind's future. He is interested in the whole company, perhaps because he is a whole man. His demands are worth the meeting, and in fairness it should be said that many writers try to give him the insights he wants. The "Operations" section is therefore a special kind of public relations document for a very important member of the public—the knowledgeable man who cannot be fed pablum. He is not naive; he is acute, sensitive to condescension, impatient of evasion.

It is comparatively easy to meet the need to know by straightforward exposition and a clear division of material. Even a maxi-conglomerate with its law suits, criminal charges, and Congressional investigations can be discussed one subsidiary at a time. The writer must, however, seek pragmatic answers to some questions. What background is he to include? The Report is after all one of a series; reports were written in preceding years. On the other hand, how many of the shareholders are new additions to the family of owners? For these the important job of public relations has still to be done. Are the Reports to be distributed to other groups—employees, customers, or possibly investors? Most importantly perhaps, in what new areas of activities is the company now engaged? Certainly they require careful exposition.

Supplying a sense of pride may be more difficult. The material available to the writer varies with each company, but it is possible to suggest some approaches. Many companies have helped in the great adventure of the century—setting Man on the Moon. Any contribution to air-space developments or even to air travel has its touch of glamor. If the company operates in remote areas of the world, the reader can take some pride in the fact that his company is helping to build, let us say, Zambia; company towns are very different from what they were when Upton Sinclair wrote of them. If his company has had part in the production of miracle drugs or miracle mechanisms, the reader is delighted.

The reader is also interested in research. Perhaps because the unknown is hard to describe, discussions of research are often vague and platitudinous. However, if such a section has a place in the Report, the writer should try to make it meaningful even when information must be withheld for proprietary reasons or reasons of national security. On the other hand, the writer must guard against over-enthusiasm about the scope, or the success, or the financial possibilities of research in progress. Excessive optimism may haunt his successors years later.

For reasons that are both practical and emotional the reader wants to know about employee relations. He is entitled to candor if they are bad; he may feel some pride in the unusual goodwill if they are good. Stockholders are also interested in the company's relations with the society, especially if most of them live within a geographically small area, for then they are being informed at once about their company and their community. In addition, events of recent years have imposed social responsibilities on the companies with which they do not yet cope successfully; if the company is making efforts to solve, say, problems of pollution, stockholders want to know about both the effort and the degrees of success.

FINANCIAL STATEMENTS

"Financial Statements" are the most solemn sections of the Report. The Auditor's Opinion, like holy oil, confers sanctity that protects them from a layman's criticism. Articles in professional accounting journals provide evidence, however, that the profession is not always satisfied with the presentation of information. The information in the Financial Statements is, in fact, not adequate for the interested stockholder. Other documents required by the Securities Exchange Commission manage to be more informative.

Although every Report tells us that the Notes are an integral part of the "Financial Statements," they are the part of the Report that is read by the smallest number of people. They are almost always so badly written as to be almost incomprehensible. This is not necessary as the very occasional Report proves. Unfortunately, they are indeed important. The stockholder must go to them to be enlightened about the significance of mergers, stock options, warrants or convertible bonds. He is not very often enlightened.

MANAGEMENT

The sparse information about company management is usually placed on the back page of the Report, frequently on the inside back cover. It is likely to consist principally of the names of the members of the board and the senior officers. Occasionally the section includes photographs of all these gentlemen, although they are not really very decorative. Simple as this section is, it can be made more useful. Unless the document is the General Motors or the A.T.&T. Report, the reader is probably interested in the business affiliations of the members of the board. He wants some information about the talents, experience or business connections that they bring to the direction of the company.

FORMAT AND PRESENTATION

To this point this discussion has been concerned with the narrative of the Report. But the effectiveness of a Report depends on many things beside its golden prose. It is possible to exploit the resources of the graphic artist, the photographer, and the printer to turn the Report into an artistic, or at least effective, document. As every report is different, there are no absolute rules, but we can speculate the reason why some devices appeal and others do not.

THEMES AND OTHER GIMMICKS

Regretfully the frequent reader of reports feels that most innovations fail. The Litton Industries Reports of 1967 and the years following are well known; they may well become collector's items. The reproductions of works of art, all related to industry and trade, make the 1967 Report striking, but it is, after all, a tour de force. The primary purpose of the Report was in no way changed. In 1966 Oak Electro-Letics prepared a package of three reports, each bound separately. One was designed for stockholders, one for customers, and the third for analysts. It seemed like a good idea; yet I believe it failed of its purpose. The package was very expensive; much material was repeated in all three; and inevitably, anyone using any of the versions found that he wanted some of the material included in the others.

Occasionally Reports (including those by Litton Industries) have been built about a theme. Again, the idea seems good, but the treatment becomes forced. What themes, after all, are there? In specific years they may be *Growth*, or *Expansion*, or *Diversification*, or *Profitability*. These are well stated in a President's Letter, but as a theme on the title page, they recall the class motto of the High School Annual—and what is to be put there next year? One insurance company selected as a theme a working day of one of its agents. It appeared that he was a combination of financial analyst, family counselor, and father confessor, all on a twenty-four hour basis. There seemed to be little time left for him to earn some money. The hard-sell promotion alienated many readers. On the other hand, one bank devoted much of its Report to a clear explanation, accompanied by cartoons, of each of the departments of a large bank. It was pleasant to read, and no reader could help understand more about this bank and about banking in general. Nevertheless, one is forced to concede that the usual format has the advantage of familiarity, and any departure from it demands very strong justification, for it makes even the willing reader a little uncomfortable.

PRETTY PICTURES AND OTHER GRAPHICS

Graphics, not surprisingly, are graphic, and one picture, we all know, is worth a thousand words. No one would suggest omitting drawings, photographs, or graphs from Annual Reports. They appeal to all levels of users; they clarify; they impress facts and relationships upon the memory. Yet they are not always as effective as they might be. Even when there is sharp photography and technical skill, there is often some floundering about what should or should not be used. If the composers of reports keep in mind the basic purposes: to increase knowledge and understanding, and to create some sense of satisfaction in the work of the company, they can make judgments about inclusions and omissions.

There is no doubt that bar diagrams, silver dollars divided into segments, and graphs help the reader. Unfortunately, they are frequently untitled, and the items that bars or axes represent are stated too cryptically. On the other hand, there are limits. Eighteen or twenty-four small graphs filling two solid pages are certain to lose many readers.

Certain devices almost always please. Maps of the world, for instance, can show the routes of airlines; they can be marked to show the areas where the company is active in sales or production. There is still something of the romantic in most people, and the feeling that through one's company one shares in activities in Bombay, Bangkok, or Bali is better than reading a short story by Somerset Maugham, for the stockholder is part of it.

Photographs of course help create this sense of excitement. Unfortunately, however, a building in Manila does look very like one in Milwaukee. A caption for rows of trucks on the docks at Haifa is not convincing if they look just like trucks on a dealer's lot in Des Moines. One looks for evidence of distant lands in costumes, or people's faces, or exotic architecture. Perhaps the least useful photographs are the very common wallet-size photographs of members of the Board, spread over two dull pages.

Pictures of products are usually interesting; there is double justification for their inclusion if the company manufactures food, clothing, or any other consumer goods. Readers want to know just what the company manufactures, and in this day of conglomerates they do not always know. In addition, pictures of products are excellent advertising; there can be no more receptive audience. A manufacturer of men's clothing who reproduces the advertisements the company ran in *Esquire* can, with reason, think of his Report as the best way to use his advertising dollar—especially since the dollar has already been spent!

THE COMPARISON TABLE

Some reports include a "Comparison" of the activities of the company over a period of years. Sometimes the comparison or review is combined with the "Highlights," and then it is likely to be very sketchy. The "Comparison" can cover two full pages and can include as many as thirty items, ranging from sales and profits to ratios to dividends, assets, numbers of employees, numbers of shareholders, branches, plants, etc. It is surprising that this feature does not appear in many more of the Reports. It cannot by itself substitute for a prose discussion of the affairs of the company, but there is no other way that so much information can be presented so directly and so clearly. It would probably increase significantly the number of those who read past the first few pages. There are strong reasons for including the section, and no reason for its omission except the possible desire to obscure unfavorable trends.

Reports differ because companies are different and report writers differ in taste and in the judgments they make. But it is possible to generalize. There is frequently no sense of direction. Reports can be made more useful if the composers keep firmly in mind the diversity of readers, their different levels of expertise, and their needs. Then writers can write for the different audiences, and make valid decisions about graphics, photographs and tables. If they help the readers to understand the affairs of his company and to empathize with the organization of which he is a tiny part, they are successful. When they merely fill space, even though they fill it attractively, they fail.

DISCUSSION AND ACTIVITY

1. As Fiskin points out, corporate annual reports have developed a format that includes Highlights, President's Letter, Operations, Financial Statements (usually followed by Notes to the Statements), and Management. Team up with a classmate, collect some sample corporate annual reports, and check how closely the sections of the reports conform to Fiskin's analysis.

2. Select a corporate annual report, study it carefully for its description of the company's finances, and write a report explaining how the writer handles good news and bad news.

Suggestions for Further Reading

APPLICATION LETTERS AND RÉSUMÉS

Issel, C. K. "The Résumé—A Sales Tool." *Technical Communication* XVIII (May/June 1971), 7–9.

The résumé should be planned like a sales document and should be limited to one page of real information that presents a summary of one's professional objective, ability, and background. "Objective" is one's goal at this point in life, not what one wants to be twenty years from now. Ability to perform in the job one is applying for is reflected by past experience. One should choose those experiences that reflect successful performance and high potential for reaching one's stated objectives. "Background" is one's education and brief personal facts about oneself. In a personal description, give date of birth, family status, height, weight, and state of health.

Weismantel, G. and J. Matley. "Is Your Résumé Junk Mail?" *Chemical Engineering* LXXXI (November 11, 1974), 164, 166, 168.

Drawing from a booklet titled *Job Hunting: The Seven Steps to Success* (published by Lockheed Aircraft Corporation), the authors suggest that when one compiles a résumé and application letter, one should apply the lessons of the direct mail business: keep the résumé short, get to the point, be specific, write a good cover letter and address it to a particular person, and be a little different (use light gray bond paper instead of white bond paper, for instance).

MANUALS

Deck, Warren H. "What Makes the Good Instruction Book?" *General Electric Review* 58 (July 1955), 40–43.

The good instruction book aids the customer in installing, operating, and maintaining equipment. It should contain adequate information in easy-to-understand format. The style should be simple, exact, and concise, and plenty of subheads should be used. The writer should give commands to the readers as if he or she were speaking directly to them.

Wright, John W. "Instruction Manuals." *Product Engineering* 24 (August 1953), 141–148.

Military and commercial manuals are the two major types of manuals. The same general techniques and requirements apply to both types. However, military manuals must follow additional specific requirements established by the military, and commercial manuals are a more elaborate presentation because they are, in part, selling tools. Manuals may be operating instructions, service instructions, overhaul instructions, or parts lists. The manual writer should use short sentences and paragraphs, give specific instructions, avoid uncommon words and unnecessary phraseology, use exact nomenclature, and use headings and illustrations frequently.

REPORTS

Brief, Arthur P. and Allan C. Filley. "Selling Proposals for a Change." *Business Horizons* 19 (April 1976), 22–25.

Knowing how to package a proposal and knowing the variables that are likely to influence a review committee will help increase the likelihood of the proposal being adopted. A proposal should show confidence that its objectives can be achieved, should be easy to understand, and show that the ratio of costs to benefits is favorable.

Houlehen, Robert. "Writing Reports that Communicate." *Industry Week* 186 (September 22, 1975), 46.

The writer's goal is to organize and simplify information that somebody else may need. The writer can save time and achieve good results if he or she knows the organizational patterns and the support possibilities for reports (general-to-particular, known-to-unknown, order of importance, time sequence, cause-and-effect, pro-and-con, and classic-persuasion).

Souther, James W. "Design That Report!" *Journal of Chemical Education* 28 (October 1951), 519–520.

Reports should be designed to fit a specific situation, purpose, and audience. The writer must consider the requirements that may be imposed upon him or her by the audience or by company policy. General conclusions, summaries, and recommendations should be placed at the beginning of the report; the more detailed information toward the end of the report. If the writer is able to analyze the writing situation and knows the writing process, he or she will find writing reports much easier.

Tutt, Charles L., Jr. "Preparation and Evaluation of an Industrial Report." *General Motors Engineering Journal* 4 (April/May/June 1957), 30–33.

Because reports assist management in making policy and decisions, they should be complete, concise, and accurate. Reports may vary in length, but they should have a uniform format of preface, introduction, conclusion, main discussion, and appendix (if necessary). A guide sheet for evaluating reports is included at the end of the article.

Part V

What Are the Important Formal Elements of Reports?

"The writer does the most who gives his reader the most knowledge and takes from him the least time."
                                    Sydney Smith

We do not like the term "formal report." We admit that some documents are called "formal reports," but we prefer to speak of *"formalized reports."* What we object to is regarding reports as either formal or informal. Such black-white, either-or categorizing simply doesn't reflect the variety of written reports used in business, industry, and government.

Formal reports are at one extreme of a continuum. What actually exists is a range of report formats running all the way from a few handwritten sentences on a note pad to longer, more permanent reports that require varying degrees of formalization. A report is formalized by adding certain elements to the basic text for the purpose of highlighting important information. The complete list of formal elements would include title page, table of contents, list of figures, letter of transmittal or preface, abstract, heading system, references, appendixes, and formal graphics.

Sometimes a report may be slightly formalized by having a title page and abstract. Other times, a report may use a title page, letter of transmittal, and abstract. Still other times, especially when the report is of permanent value and intended to serve as a long-time reference, it may use the full complement of formal elements. Many times, you will have to conform to your organization's style manual or to your customer's specifications concerning format of reports. When you have no particular guides to follow, you will have to decide which formal elements to incorporate into your report.

Let us spend a little time examining the need and method for formalizing reports.

As you have seen earlier in this book, your main concern as a writer is to give your reader essential information as conveniently as possible. Thus, it is of utmost importance that you concentrate on getting through to the reader in a way that makes it easy for him or her to read your entire report, to find particular parts of your report, or to retrieve data from your report. In Part II you saw how a clear, economical style can make your writing more easily understandable. In Part III you saw how certain writing strategies can make your writing engage the reader. In Part IV you learned how patterning your writing in expected, standardized ways helps the reader follow your presentation. The selections in this fifth section explain how formalizing a report helps the reader get an overview of your report and find what he or she needs from your report.

An organization that requires its individuals and committees to write and read many reports will frequently establish guidelines for formalizing reports. The purpose of such formalization, as you now recognize, is to speed up the access of information by building redundant features into the report. A redundancy is anything that repeats itself. If the redundancy is unnecessarily repetitive, it is superfluous—bordering on absurdity—and wastes the reader's time. But if the redundancy is purposefully built in to assure that no failure of communication occurs, then it is useful. Reports longer than three or four pages must make their subject, purpose, scope, conclusions, organization, and supporting data prominent.

The title and abstract of formalized reports are so important in helping readers decide whether to scan or read thoroughly that both John D. Stevens's "Polish Your Title and Abstract" and Christian K. Arnold's "The Writing of Abstracts" have been included. Reports cannot be stored, retrieved, or read handily without good titles and abstracts.

Perhaps nothing makes a reader's heart sink more quickly than a report that consists of pages of text unbroken except for paragraphing. A system of headings and subheadings will highlight the important data of your report by lifting key words and phrases from the text and setting them off by white space, underlining, or different type face. J. Raleigh Nelson's "Sectional Headings as Evidence of Design" discusses this important formalizing element. However, since the format of headings and subheadings is not standardized, you will see several slightly different formats in reports you read.

Formalizing reports requires special attention to format—the physical placement of the report on sheets of paper. Anne Shelby's "How to Type Your Paper" explains an easy-to-follow procedure for typing a formalized report: the title page, the table of contents, the body of the report (including heading system, quotations, and footnotes), and bibliography.

Selection Twenty-Five

Polish Your Title and Abstract

JOHN D. STEVENS

Stevens suggests that writing your abstract before you begin the body of your report will serve better than an outline to keep you on your subject. He believes that you should make the title of your report tell something about the nature of the study and suggest the findings.

Right now, you are wondering if this article will be worth your time and effort to read. Does it contain any new ideas? Does it have to be read now, or can it go on the "one-of-these-days" stack?

Your reader asks the same questions when he picks up your technical report. He will base his answers on the title and the abstract. That's why you had better give those elements some careful consideration. If your report is not read, you are not communicating.

The popular writer—the novelist, the magazine writer, the news-paperman—knows he has to capture the reader's interest and then use every devise at his command to hold it. The first time you let up, he will be gone.

It is time technical writers acknowledged this. Many of them will spend months on the body and dash off the opening sections. Still, hundreds or even thousands of readers are going to see the title and/or abstract for every one who wades through the entire report.

Abstracts are a reality of modern professional life. No one has time to read all he should, let alone what he would like. It would take one man 465 years to scan—not read—the 60 million pages of technical reports produced in the world during 1960. In 24 hours, enough are turned out to fill seven sets of the Encyclopedia Britannica.

Abstracts are carried in three places: abstract journals, library index cards,

Rubber Age 89 (September 1961), 988. Reprinted by permission of the publisher.

and preceding many journal articles. In each place, they are scanned by many readers.

The typical scientist or engineer reads a handful of pet journals and relies on abstract journals to fill him in on material published in journals in his own and related fields, and to help him spot articles he should read in full.

These journals (and there are about 300 of them covering scientific fields now) do such a good job of summarizing significant results that it usually is not necessary for the reader to go through the entire report—unless he is particularly interested in details about procedures.

THE ABSTRACT

Write your abstract before you begin the body of your report. It will serve better than an outline to keep you within the bounds of your subject matter.

Benjamin Franklin reportedly finished a letter to a friend with the comment, "I hope you will excuse this long letter; I didn't have time to write a short one."

Concise writing is more difficult, but boiling down information—*in advance*—will help you when you write the full-length material.

Here is a formula for writing abstracts which may prove helpful.

State the problem in a single sentence. This will take some time and effort, but it is the key to the abstract and to the report. Make sure it really states the problem. Then start cutting down the sentence. Become a word miser. Imagine you are an editor with a terrific space shortage (because this is exactly the situation every editor faces). Is every word essential?

Once you get the subject sentence trimmed to the bone, write two or three more sentences expanding the explanation. Here you may be able to put back some material boiled out previously.

Now state your approach in a single sentence. Follow the procedure you used to state the problem. Cut the sentence; then write two or three more supporting statements. Then cut those.

Now you are ready to state the really significant results. Decide which results belong in the abstract. Seldom is there justification for listing more than one. Support it with a couple of sentences.

If you limit the abstract to 250 words (about one double-spaced typewritten page), you will be turning out really meaty abstracts—the only acceptable kind.

We propose a stiff test for any abstract. Hand just the abstract to two or three friends who have college degrees in other fields. After they have read it, ask them to tell you the problem in their own words; then, the approach; then, the significant results. They may not understand the formulas or the technical terms, but they should have some idea of what the report is about.

Technical writers must never forget that most of their readers are not experts on the subject. If they were they would be doing the writing. Most research reports are directed to management personnel who have little or no technical background. They basically are businessmen.

WRITING THE TITLE

Now, let's consider the title. The title should be complete and easy to understand. Above all else, it should identify the document. Conciseness is desirable.

Keep in mind that the report or article is going to be indexed and listed according to this title, so make sure the person (or machine) doing the filing *cannot* misfile it. If she (or it) does, the information is lost.

The title is more than a label. It should tell the reader something about the exact nature of the study, and if possible, it should suggest the findings. It should do all this in eight words or fewer.

For instance, "A Report on Some Aspects of Growth Factors in the Puerto Rican Rubber Industry" is about as bad a title as one could write. "Growth of the Puerto Rican Rubber Industry" tells the reader just as much in fewer words, but can be improved.

A good, strong verb almost always livens up a title. The active voice gives your title a sound of authority.

"Rubber Industry Triples in Puerto Rico" makes a positive statement, and gives the reader the gist of the findings.

Another good approach is the "How" title. This subject might be treated as "How Puerto Rico Tripled Rubber Production." But before using the "How" title make sure your report answers the question!

There is no excuse for such musty-sounding terms as "A Report on," "Some Aspects of," or "Some Observations on." The reader can see it is a report, and he assumes it covers "aspects" and "observations."

The title is the key to the article or report. If it does not attract prospective readers, the rest of the report might as well not be there.

Once the title has attracted the reader, it is up to the abstract to hold him long enough to provide the really pertinent facts. If both parts do their job, the reader will know whether he needs to proceed.

Both the title and the abstract are worthy of the author's careful consideration. It is time technical report writers stopped doing these important elements as afterthoughts and started treating them as the really meaningful parts of the report.

Selection Twenty-Six

The Writing of Abstracts

CHRISTIAN K. ARNOLD

Previously an instructor of English and technical writer and editor, Arnold is associate executive secretary of the Association of State Universities and Land-Grant Colleges. Arnold points out that effective abstracts must (1) contain adequate information to satisfy the needs of a researcher looking for sources of information and of an administrator looking for a progress of status definition; (2) be a complete, self-contained unit; (3) be as short as possible without eliminating essential information; (4) be written in fluent, easily understood language; (5) be consistent in tone and emphasis with the parent paper or report; and (6) make the widest possible use of numerals and standard, generally recognizable abbreviations.

The most important section of your technical report or paper is the abstract. Some people will read your report from cover to cover; others will skim many parts, reading carefully only those parts that interest them for one reason or another; some will read only the introduction, results, and conclusions; but everyone who picks it up will read the abstract. In fact, the percentage of those who read beyond the abstract is probably related directly to the skill with which the abstract is written. The first significant impression of your report is formed on the reader's mind by the abstract; and the sympathy with which it is read, if it is read at all, is often determined by this first impression. Further, the people your organization wants most to impress with your report are the very people who will probably read no more than the abstract and certainly no more than the abstract, introduction, conclusions, and recommendations. And the people you should want most to read your paper are the ones for whose free time you have the most competition.

Despite its importance, you are apt to throw your abstract together as fast as possible. Its construction is the last step of an arduous job that you would rather have avoided in the first place. It's a real relief to be rid of the thing, and almost anything will satisfy you. But a little time spent in learning the "rules" that govern the construction of good abstracts and in practicing how to apply them will pay material dividends to both you and your organization.

The abstract—or summary, foreword, or whatever you call the initial thumbnail sketch of your report or paper—has two purposes: (1) it provides the specialist in the field with enough information about the report to permit him to decide whether he could read it with profit, and (2) it provides the administrator or executive with enough knowledge about what has been done in the study or project and with what results to satisfy most of his administrative needs.

It might seem that the design specifications would depend upon the purpose for which the abstract is written. To satisfy the first purpose, for instance, the abstract needs only to give an accurate indication of the subject matter and scope of the study; but, to satisfy the second, the abstract must summarize the results and conclusions and give enough background information to make the results understandable. The abstract designed for the purpose can tolerate any technical language or symbolic shortcuts understood at large by the subject-matter group; the abstract designed for the second purpose should contain no terms not generally understood in a semitechnical community. The abstract for the first purpose is called a *descriptive abstract*; that for the second, an *informative abstract*.

The following abstract, prepared by a professional technical abstracter in the Library of Congress, clearly gives the subject-matter specialist all the help he needs to decide whether he should read the article it describes:

Results are presented of a series of cold-room tests on a Dodge diesel engine to determine the effects on starting time of 1) fuel quantity delivered at cranking speed and 2) type of fuel-injection pump used. The tests were made at a temperature of −10°F with engine and accessories chilled at −10°F at least 8 hours before starting.

Regardless of however useful this abstract might be on a library card or in an index or an annotated bibliography, it does not give an executive enough information. Nor does it encourage everyone to read the article. In fact, this abstract is useless to everyone except the specialist looking for sources of information. The descriptive abstract, in other words, cannot satisfy the requirements of the informative abstract.

But is the reverse also true? Let's have a look at an informative abstract written for the same article:

A series of tests was made to determine the effect on diesel-engine starting characteristics at low temperatures of 1) the amount of fuel injected and 2) the type of injection pump used. All tests were conducted in a cold room maintained at $-10°F$ on a commercial Dodge engine. The engine and all accessories were "cold-soaked" in the test chamber for at least 8 hours before each test. Best starting was obtained with 116 cu mm of fuel, 85 per cent more than that required for maximum power. Very poor starting was obtained with the lean setting of 34.7 cu mm. Tests with two different pumps indicated that, for best starting characteristics, the pump must deliver fuel evenly to all cylinders even at low cranking speeds so that each cylinder contributes its maximum potential power.

This abstract is not perfect. With just a few more words, for instance, the abstracter could have clarified the data about the amount of fuel delivered; do the figures give flow rates (what is the unit of time?) or total amount of fuel injected (over how long a period?)? He could easily have defined "best" starting. He could have been more specific about at least the more satisfactory type of pump: what is the type that delivers the fuel more evenly? Clarification of these points would not have increased the length of the abstract significantly.

The important point, however, is not the deficiencies of the illustration. In fact, it is almost impossible to find a perfect, or even near perfect, abstract, quite possibly because the abstract is the most difficult part of the report to write. This difficulty stems from the severe limitations imposed on its length, its importance to the over-all acceptance of the report or paper, and, with informative abstracts, the requirement for simplicity and general understandability.

The important point, rather, is that the informative abstract gives everything that is included in the descriptive one. The informative abstract, that is, satisfies not only its own purpose but also that of the descriptive abstract. Since values are obtained from the informative abstract that are not obtained from the descriptive, it is almost always worth while to take the extra time and effort necessary to produce a good informative abstract for your report or memo. Viewed from the standpoint of either the total time and effort expended on the writing job as a whole or the extra benefits that accrue to you and your organization, the additional effort is inconsequential.

It is impossible to lay down guidelines that will lead always to the construction of an effective abstract, simply because each reporting job, and consequently each abstract, is unique. However, general "rules" can be established that, if practiced conscientiously and applied intelligently, will eliminate most of the bugs from your abstracts.

1. *Your abstract must include enough* specific *information about the project or study to satisfy most of the administrative needs of a busy executive.* This means that the more important results, conclusions, and recommendations, together with

enough additional information to make them understandable, must be included. This additional information will most certainly include an accurate statement of the problem and the limitations placed on it. It will probably include an interpretation of the results and the principal facts upon which the analysis was made, along with an indication of how they were obtained. Again, *specific* information must be given. One of the most common faults of abstracts in technical reports is that the information given is too general to be useful.

2. *Your abstract must be a self-contained unit, a complete report-in-miniature.* Sooner or later, most abstracts are separated from the parent report, and the abstract that cannot stand on its own feet independently must then either be rewritten or will fail to perform its job. And the rewriting, if it is done, will be done by someone not nearly as sympathetic with your study as you are. Even if it is not separated from the report, the abstract must be written as a complete, independent unit if it is to be of the most help possible to the executive. This rule automatically eliminates the common deadwood phrases like "this report contains . . ." or "this is a report on . . ." that clutter up many abstracts. It also eliminates all references to sections, figures, tables, or anything else contained in the report proper.

3. *Your abstract must be short.* Length in an abstract defeats every purpose for which it is written. However, no one can tell you just how short it must be. Some authorities have attempted to establish arbitrary lengths, usually in terms of a certain percentage of the report, the figure given normally falling between three and ten percent. Such artificial guides are unrealistic. The abstract for a 30-page report must necessarily be longer, percentagewise, than the abstract for a 300-page report, since there is certain basic information that must be given regardless of the length of the report. In addition, the information given in some reports can be summarized much more briefly than can that given in other reports of the same over-all dimensions. Definite advantages, psychological as well as material, are obtained if the abstract is short enough to be printed entirely on one page so that the reader doesn't even have to turn a page to get the total picture that the abstract provides. Certainly, it should be no longer than the interest span of an only mildly interested and very busy executive. About the best practical advice that can be given in a vacuum is to make your abstract as short as possible without cutting out essential information or doing violence to its accuracy. With practice, you might be surprised to learn how much information you can crowd into a few words. It helps, too, to learn to blue-pencil unessential information. It is perhaps important to document that "a meeting was held at the Bureau of Ordnance on Tuesday, October 3, 1961, at 2:30 P.M." somewhere, but such information is just excess baggage in your abstract: it helps neither the research worker looking for source material nor the administrator

looking for a status or information summary. Someone is supposed to have once said, "I would have written a shorter letter if I had had more time." Take the time to make your abstracts shorter; the results are worth it. But be careful not to distort the facts in the condensing.

4. *Your abstract must be written in fluent, easy-to-read prose.* The odds are heavily against your reader's being an expert in the subject covered by your report or paper. In fact, the odds that he is an expert in your field are probably no greater than the odds that he has only a smattering of training in any technical or scientific discipline. And even if he were perfectly capable of following the most obscure, tortured technical jargon, he will appreciate your sparing him the necessity for doing it. T. O. Richards, head of the Laboratory Control Department, and R. A. Richardson, head of the Technical Data Department, both of the General Motors Corporation, have written that their experience shows the abstract cannot be made too elementary: "We never had [an abstract] . . . in which the explanations and terms were too simple." This requirement immediately eliminates the "telegraphic" writing often found in abstracts. Save footage by sound practices of economy and not by cutting out the articles and the transitional devices needed for smoothness and fluency. It also eliminates those obscure terms that you defend on the basis of "that's the way it's always said."

5. *Your abstract must be consistent in tone and emphases with the report proper, but it does not need to follow the arrangement, wording, or proportion of the original.* Data, information, and ideas introduced into the abstract must also appear in the report or paper. And they must appear with the same emphases. A conclusion or recommendation that is qualified in the report proper must not turn up without the qualification in the abstract. After all, someone might read both the abstract and the report. If this reader spots an inconsistency or is confused, you've lost a reader.

6. *Your abstract should make the widest possible use of abbreviations and numerals, but it must not contain any tables or illustrations.* Because of the space limitations imposed upon abstracts, the rules governing the use of abbreviations and numerals are relaxed for it. In fact, all figures except those standing at the beginning of sentences should be written as numerals, and all abbreviations generally accepted by such standard sources as the American Standards Association and "Webster's Dictionary" should be used.

By now you must surely see why the abstract is the toughest part of your report to write. A good abstract is well worth the time and effort necessary to write it and is one of the most important parts of your report. And abstract writing probably contributes more to the acquisition of sound expository skills than does any other prose discipline.

Selection Twenty-Seven

Sectional Headings as Evidence of Design

J. RALEIGH NELSON

One of the first generation teachers of report writing, J. Raleigh Nelson was professor of English in the College of Engineering at the University of Michigan. His article explains how sectional headings in reports expedite reading and subsequent use of reports.

The paragraph, as has been shown, is the most useful and natural means a writer has for making the structural pattern of a composition evident as a guide to the reader. He should think of it always as a means of marking the analysis and organization of his material made in the interests of utilizing the attention of his reader most economically and effectively. He should diversify his transitional devices till his finished work shows a coherent handling of paragraphs which is so natural and unobtrusive that the reader is quite unaware of how it has been effected.

As evidence of organization, no substitute can be as satisfactory as an orderly, coherent paragraph scheme. In reports, however, where everything must be done to expedite both a rapid first reading and a ready use of the text for frequent reference reading, there have developed in the course of years other still more obvious means of marking the plan of organization. It was stated . . . [before] that one way of keeping the reader from getting lost in the maze of detail in the body of one's exposition is to mark the unit parts distinctly so that he is always aware of what is *a* and what is *b* and what is the point at which he passes from *a* to *b*. It has been shown that the topic sentences, with the various explicit transition words and phrases and transition sentences, or paragraphs as the case demands, are intended to do just that. But in a report the route must be charted even more obviously. There is need for additional safeguards to assure even the most cursory reader against confusion. Headings marking the various sections are such supplementary means—almost mechanical and quite foolproof. It might be said, rather crudely, that in a report, the unit sections are taken apart and actually labeled for the reader's convenience. One might, perhaps still more crudely, say that the reader of a report is provided with road signs; in the headings he has a means of making his way through the report without getting lost and with the assurance that he will reach his destination safely.

There has been, and still is, no perfect agreement as to any one system of headings. It is a usage that has not yet been entirely standardized, although each year has seen a nearer approach to uniformity; certainly there is beginning to be a more conscious use of such headings as a means of marking the plan of a report; they are no longer used, as often was the case years ago, merely to call attention to some one part that needed emphasis. The system that this book suggests seems soundly based on the theory that the headings have value only as they help the reader find his way about in a report and expedite his handling of it. It seems to meet the requirements for a system that can always be relied on to represent the structural plan of the report, which agrees perfectly with the table of contents, and which indicates always and consistently the relation of main and subordinate units in such a plan. Moreover, it is flexible enough so that it can be applied, according to one's best judgment and experience, as it is needed. One section, where the coherent flow of ideas must be uninterrupted, may be marked by a single main heading, whereas in another section of the same report, where it is important to stress sharply the subordinate units, second- or even third-order headings may be used. It is the system that the author employed in editing the hundreds of typed reports that passed through his office monthly during his years as editor of the Department of Engineering Research and with slight modification to the published bulletins and circulars. It seems consistent and practical.

The specifications for this system as applied to the typing of both the long-form and the short-form report should be studied with reference to the illustrative examples that follow:

1. In the long-form report the main headings are displayed in capital letters in the center of the page. They are underscored to mark them as distinct from the text. In single-spaced text, three spaces before and two after the headings, in a double-spaced text, four spaces before and three after are recommended.

2. The second-order headings in such reports are displayed at the margin in a line above the text they introduce. Capital and lower-case letters are used, and the headings are underscored. A period usually follows such headings. The text begins on the line below. The spacing is increased by one space preceding the heading but is kept the same following it.

3. The third-order heading in the long-form reports is like the second-order; it is displayed in capital and lower-case letters underscored. As an inset at the beginning of the first paragraph in the section it marks and is followed by a colon and a double space preceding the text, which follows in the same line with the heading.

4. The short-form reports omit from the series of headings the heavy, all-capital headings; they use the second-order headings as main headings and the third-order headings of the long-form reports as their second-order headings. Short-form reports are usually single-spaced.

5. The headings, the outline, and the table of contents should agree as to relation of parts.

Example 1. *Two Orders of Headings in a Single-spaced Long-form Report.*

Main heading   PRESENT PRINTING SYSTEM

                After the character of the work has been simplified and made as uniform as possible by means of the foregoing suggestions, the printing department itself should be considered with the end in view of making any necessary changes to cut the cost of producing that work. In order that the reader may have a clear picture of the routine of the plant as it now operates, a brief description is necessary.

Second-order   Job Ticket.
heading   The first step in the execution of a printing order is the making out of the job ticket. These are made from requisitions coming from the Purchasing Department, through which office all orders in the university must come. The job ticket itself consists of a manila envelope, printed on one side with a form calling for all information concerning the job and with sufficient room inside for all samples, proofs, and special instruments which should be kept with the job. This job envelope, or "ticket" as it is usually called, never leaves the job but follows it through the shop to completion. When the job is completed, the ticket is taken into the office; and after certain information required for billing the job is taken from it, it is filed.

Second-order   Composition.
heading   With the ticket made out and the samples of printed material inside it, the job is now ready for composition, the first operation in printing.

The office manager takes the ticket to the
compositor who, after inspecting the samples,
determines from his files whether or not the form
is standing.

Example 2. Second- and Third-order Headings in a Single-spaced Long-form Report.

*Second-order
heading*
*A good
example of a
transition and
introductory
paragraph to a
section*

Profitable Enterprise.
  So far we have not considered the opportunities
that the present situation presents, and we have
been inclined to stress the dependence of these
opportunities on the solution of the problem of
legislation. Yet past experience shows that, even
without such legislation, flying operations of a
commercial nature may be carried on with profit.
Already, a variety of enterprises have proved
successful. These include not only passenger and
freight carrying, but such specialized work as
photographing and aerial advertising by what is
known as "sky-writing."

*Third-order
heading*

Passenger Carrying: The first field of aerial
activity capable of commercial exploitation was
that of carrying passengers on short flights. This
enterprise found immediate favor at the close of
the war and offered an opportunity to the
individual flying his own plane as well as to the
organized company. It served also to acquaint more
people with aircraft and helped to break down the
attitude of skepticism and prejudice. This field
still offers considerable opportunity, and it will
be found that a reputable concern operating near a
large city can do $300 or $400 worth of business on
Sundays and holidays.

*Third-order
heading*

Freight Carrying: As the gradual demands for
passenger service between cities and towns led
companies of a more permanent nature to establish
scheduled intercity service, it was natural that
freight transportation should enter in as an
auxiliary. The expense connected with such
shipments limited the scope of this work, and in
general, it was found that only with goods of
perishable nature was the gain in speed able to
make up for the advance in cost.

During the past year, however, the newspapers have
begun to use aircraft for rapid transmission of

photographs of exceptional events. By means of aircraft, photographs of the Dempsey—Firpo prize fight which left New York shortly before midnight were in Chicago in time for the morning editions, and similar service has been rendered for other prize fights and such national calamities as the earthquake in Japan.

At present, this field offers but few, and irregular, opportunities; but there are many indications that it may yet assume the proportions of a major enterprise and is one that is capable of yielding large profits.

Example 3. Sketch of an Entire Short-form Report, Showing the Slightly Different Use of Headings.

OPERATING COSTS FOR AIR—CONDITIONING SYSTEMS USING SILICA GEL AND ELECTRIC REFRIGERATION

Air—conditioning systems using silica gel, electric refrigeration, or both together are compared in this report to show the relative operating costs of each with different types of cooling loads. The report was requested by Mr. J. H. Walker to set forth the effect of moisture removal on the operating costs of air conditioning with different types of systems. Four different systems are considered. These are, a commercial silica gel unit using deep—well water for cooling, a modified silica gel unit using city water and rehumidification for cooling, an electric—refrigeration unit using city water for condensing, and a combination electric—refrigeration and silica gel unit using city water and heat to reactivation of the gel beds for condensing. These units are outlined in the diagrams on Plate 1. They are compared with particular regard to costs and conditions in the residential area of metropolitan Detroit.

Effect of the Moisture Load.
A discussion of the effect of the moisture load may be helpful in correlating the cost characteristics of the different systems. [This topic is developed in seven paragraphs.]

Assumed Operating Conditions.
Since the degree of comfort in a room is determined by the effective temperature, the four cases considered are computed for a common effective temperature of 71° Fahrenheit. [There follow three paragraphs.]

Conclusions.

The operating costs shown on Plate 2 and previously
summarized indicate that the units using silica gel do not
have much to recommend them. Their operating costs are
higher in most cases and they are probably more expensive
than an electric-refrigeration unit. The combination unit,
which has the most favorable operating cost, is certainly
much more expensive. Simplicity in the control mechanism
also should not be overlooked. The electric-refrigeration
system has fewer distinct processes to control and would
therefore have a much simpler and more dependable control
mechanism. Of the four cases considered, it is probable that
the electric-refrigeration unit offers the lowest over-all
cost for air conditioning. This conclusion is only slightly
affected by the magnitude of the moisture load as may be seen
from Plate 2.

Example 4. Sketch Showing the Relation of Headings to the Outline Plan.

THE USE OF HEADINGS IN DOUBLE-SPACED TEXT

I. Introduction
II. Development of Welding Unit
    a. Division of piping system into suitable units
    b. Shop welding of pipe to valves and fittings
       1. Preparation of valves for welding
       2.
       3.
       4.
    c.
       1.
       2.
III. Protection of Field-welded Joints
    a.
    b.
    c.
IV. Summary

THE USE OF HEADINGS IN DOUBLE-SPACED TEXT

This sketch will illustrate the use of the headings through
part of the first section of the student long-form report
outlined above. The text of the introduction will follow the
display of the title on the first page without the display of
the word "Introduction'' as a heading; such a display is
superfluous. Four to six spaces will be allowed between the
title and the text.

DEVELOPMENT OF WELDING UNIT

This is the first main heading. Four spaces precede this display, and three follow the text it introduces. It is in capital letters followed by no period; it is underscored.

Division of Welding Unit.
This is a second-order heading. Capital and lower-case letters are used as shown, followed by a period; it is underscored; three spaces before and two after this heading. The text follows, as in this example, on the line two spaces below the heading.

Shop Welding of Pipe into Suitable Units.
Text in the line below, as in the paragraph above.
Preparation of Valves for Welding: This is a third-order heading. It is displayed in capital and lower-case letters, as shown, aligned with the paragraph indentions. It is followed by a colon, and, after 1 space, by the text in the same line.

Example 5. Sketch Showing the Relation of the Headings to the Table of Contents

TABLE OF CONTENTS

Page

Introduction
Development of Specifications
    Simple Rules
    Semicomplete
    Complete
        Material Tests
        Field Trials
        Field Recommendations
        Standardizing of Material
        Approval
Training of Users
    Division Foremen
    Crew Foremen
    Field Estimators
    Planning Engineers
Results
    Adequate Strength

Note: The introductory text will follow the display of the title, without any heading marking it. Its introductory character is sufficiently evident from its position preceding the first main heading, which marks the beginning of the body of the report.

Reduced Stock
Required Clearance
Improved Appearance
Lower Cost
Summary

DEVELOPMENT OF SPECIFICATIONS

Simple Rules.
Semicomplete.
Complete.
  Material Tests:
  Field Trials:
  Field Recommendations:
  Standardizing of Material:
  Approval:

TRAINING OF USERS

Division Foremen:
Crew Foremen:
Field Engineers.

RESULTS

Adequate Strength.
Reduced Stock.
Required Clearance.
Improved Appearance.
Lower Cost.

SUMMARY

Note: The headings in this sketch are displayed in the exact relation to each other and are placed on the page in the precise way they will be used in the text. One will be saved many a slip if he will translate his outline plan in the form just shown on page 275 to that shown on this page.

Selection Twenty-Eight

How to Type Your Paper

ANNE SHELBY

> *Anne Shelby teaches at Hazard (Kentucky) Community College. Her article provides easy-to-follow instructions for typing a formalized report.*

If you can't find—or can't afford—anybody to type your paper for you, you'd better learn how to do it yourself. It isn't hard, but it does take time and patience. Whether you type 80 words a minute or hunt-and-peck 80 words an hour doesn't really matter once you get the paper typed. What matters then is how it looks—if it's neat and easy to read, and if it follows the rules for typing a paper.

Those rules are important (you'll have to display your individual creativity some other way). The most intellectual of college professors can be quite petty about things like margins and footnotes. They tend to frown on typing that breaks the rules, and smile on papers that look good. If you get specific instructions for typing your paper, follow those. If you don't, you should be safe in using the instructions given here.

There are three big steps: (1) getting ready to type, (2) typing the body of the paper, and (3) typing the footnote section (if you have one), the bibliography, the title page, and the table of contents.

GETTING READY TO TYPE

Before you sit down to type, make sure you're ready. That means having the right materials, making a guide sheet, and setting up the typewriter.

The Technical Writing Teacher I (Winter 1974), 11–22. Reprinted by permission of the author.

WHAT TO GET

You'll need a typewriter that works, paper, an eraser, and a ruler.

THE TYPEWRITER   Use the same typewriter for the whole paper, and be sure it works well enough to get you through. Pica type (ten spaces to the inch) or elite type (twelve spaces to the inch) is preferred over script, italics, or other unusual types. Make sure that the tab, margin, and space settings work, that the keys are clean, and that the ribbon prints clearly (always use black).

PAPER   Use 8½ x 11 inch white paper with at least 25 percent rag content. You'll probably have to retype some pages, so get a good supply (so you won't be running around in the middle of the night looking for a piece of bond).

ERASER AND RULER   Get an eraser made especially for erasing typewritten manuscripts, the hard, grainy, white kind. You'll need a ruler for measuring off the margins.

WHAT TO DO

Before you start typing, make a guide sheet and set up the typewriter.

MAKE A GUIDE SHEET   You'll need to know how close you're getting to the bottom of the page as you type, so you can have a one-inch margin at the bottom, and so you can have room for footnotes if you're going to put them at the bottom of the page (more about this later). So you make a guide sheet. On a clean sheet of paper, type the numeral "1" at the right edge one inch up from the bottom. On the single space above the "1," type "2," "3," "4," and so on, up to about the middle of the page (see Figure 28-1). Whenever you start a new page, put the guide sheet behind it, and a little to the right so you can see the numbers. Those numbers tell you how many more lines you can type on that page. When the "1" pops up, start a new page. It's also a good idea to draw a straight line from the top to the bottom of the guide sheet, one inch from the right edge, so you can tell at a glance where the right margin is. Your finished guide sheet should look like that shown in Figure 28-1.

SET THE TYPEWRITER   Set the left margin 1½ inches from the left side of the paper (15 pica spaces, 18 elite). Set the right margin one inch from the right side of the paper. You'll also have one-inch margins at the top and bottom. For the top one, figure out how many vertical spaces make an inch on your typewriter (it varies), and space down that number at the top of each page that doesn't start a major section—those will start two inches from the top.

Clear all the old tabs on your machine. Set a tab eight letter spaces to the

Figure 28-1. Guide sheet

right of your left margin for paragraphs, footnotes, and single-spaced quotations. Set a tab 4½ inches from the left edge of the paper for centering.

The text of your paper should be double-spaced. Footnotes and long quotations are single-spaced, and triple spacing will follow the headings. For now, set the line spacer for double spacing.

TYPING THE BODY OF YOUR PAPER

The procedures for typing each page of the body of your paper—from page 1 of the text to the last page of the conclusion—will be pretty much the same, so you'll need to know about all the procedures before you start. Then you can go back to the section you need as you meet each problem. The things you'll need to know about are (1) headings, (2) footnotes, (3) quotations, (4) word division, (5) page numbers, and (6) corrections.

HEADINGS

Each of the three kinds of headings is typed a different way. In order of importance they are (1) major headings, (2) side headings, and (3) paragraph headings. Figure 28-2 shows what they look like.

MAJOR HEADINGS   Each major heading of your paper—like a chapter title— should start on a new page. Space down two inches from the top, and center the heading between the margins. To center, go to the center tab and backspace half the number of letters in the major heading. Then type the heading from there in all caps. Use no end punctuation unless the heading is a question. If the heading is too long for one line, divide it into two or three single-spaced lines, so that the top line is the longest, and each following line is shorter than the one above it (inverted pyramid style). Triple space after each major heading.

SIDE HEADINGS   Triple space before side headings. Type them against the left margin on a line by themselves. Capitalize only the first letter of the first word and of all important words. Underline the heading; double space after it; no end punctuation.

PARAGRAPH HEADINGS   Paragraph headings begin at the tab setting for paragraphs. Underline them, put a period after them, and begin the text on the same line.

could have
This section includes:

brief introduction

MAJOR HEADING

After the major heading should come a brief introduction
to give an overview of that section. The overview should parti-
tion the major section into its parts; in other words, it should
tell what the side headings are going to be.

Side Heading

Side headings, which compare to the capital letter entries
in outlines, are the subdivision after major headings. If the
material under the side heading is further subdivided, then a
short introductory paragraph should follow the subheading to
give an overview of the paragraph headings.

Paragraph Heading. Paragraph headings subdivide the material
under the side headings, and compare to the Arabic numerals of
an outline. There should, of course, be more than one paragraph heading under a side heading, and more than one side heading
under a major heading.

Paragraph Heading. The paragraph heading begins at the
regular tab setting for paragraphs. Important words are capitalized. The heading is underlined, followed by a period, and the
text begins on the same line.

1

Figure 28-2.   Heading system

FOOTNOTES

There are two places to put footnotes, either all in their own separate section at the end of the paper, or at the bottom of each page of the text. The first way is a lot easier, but the second way is often required. I'll tell you how to do both. First, the easy way.

IN A FOOTNOTE SECTION   If you put them all in the back, all you have to do as you type the text is to number them, consecutively, throughout the paper. To number them, turn the carriage toward you half a vertical space and type the number. (Like this[3].) Don't skip a space between the last word or punctuation mark of the words being footnoted and the footnote number. Keep careful track of those footnotes or you'll have a big mess when you get to the back. Figure 28-3 shows how the numbered footnote looks in the text. We'll talk about typing the footnote page later, since it will come after the text.

AT THE BOTTOM OF THE PAGE   Footnotes are always numbered in the text the same way (see paragraph above). The difference here is that the references cited on each page are put at the bottom of that page. Since each page may have a different number of footnotes, and since footnotes vary in length, the problem is leaving the right amount of space for them at the bottom of the page.

As you type the text, keep in mind how many footnotes you have on that page, and keep an eye on the guide sheet. Remember that you must have an inch of margin at the bottom after typing the footnotes. Most complete footnote references take three lines; subsequent references to the same source should take only one line. Footnotes are single spaced with a double space between each one. You'll have to estimate the length of your footnotes, count the footnotes, and look at your guide sheet to know when to stop typing text and start making footnotes.

When that time comes, type a solid line a single space below the text. The line, starting at the left margin, should be 1½ inches long: 15 pica spaces, 18 elite. Double space after the line, tab to the paragraph setting, turn the carriage toward you half a space, and type the number of the first footnote on that page. Single space the footnotes, double space between them, and observe the margins. Figure 28-3 shows footnotes typed at the bottom of the page. Check your English composition book or a style manual if you aren't sure about footnote forms.

QUOTATIONS

Short quotations are simply "run in" with the rest of the text and enclosed in quotation marks, as in the first paragraph of Figure 28-4. But direct quota-

9

at random sampling are seldom made.[7] The use of control groups
in ABE follow-up studies is also rare (as evidenced by this
investigator's extensive but futile search for even one study
utilizing a control group).

Roomkin asserts that control groups can and should be
employed in evaluating ABE programs, but recognizes the problems
which, according to other researchers, make the establishment
of an adequate control group impossible.[8] These researchers
claim that findings in ABE follow-up studies with control groups
may not be valid because of the difficulty in establishing a
control group like the experimental group in all aspects except
exposure to adult basic education. To assign undereducated
adults to either participate or not to participate in ABE appears
to most researchers to be, at best, impractical. Those adults
who decide to participate in ABE are believed to have characteristics which set them apart from their undereducated counterparts
who do not elect to participate.[9] Differentiating traits like
motivation, self-confidence, and belief in education would likely
to bias the results of the study, so it would be difficult to
determine the extent to which long-range differences between
those who did and did not receive the treatment (ABE) were attri-

---

[7]Myron Roomkin, "Evaluating Basic Education for Adults:
Some Economic and Methodological Considerations," Adult Education
XXIII, 1 (1972), 26.

[8]Roomkin, p. 28.

[9]Richard Malcolm, West Virginia Long-Range Follow-up Study
(Morehead, Ky.: Morehead State University, August 31, 1970), p. 7.

Figure 28-3.   Footnote numbers in text and footnotes at bottom of page

11

Researchers seem to agree that the questionnaire should be as simple as possible. McKinney and Oglesby suggest testing the questionnaire for readability level "to insure that the content is pitched neither too high nor too low for the intellectual capabilities of the respondents."[13]

Recent studies show that the questionnaire should be personalized, as well as simple. Every attempt must be made to get responses from all the subjects.

> In a Wisconsin study with a 46.9 percent response, it was discovered that 78% of the former students in the top percentile of their class returned the questionnaire, while only 28% from the bottom percentile returned the questionnaire.[14]

The tendency for successful students to return more questionnaires than their less successful counterparts can invalidate findings and conclusions.

W. R. Snelling suggests that a high response rate is possible from all segments of the population if the questionnaire is highly personalized. Snelling hypothesized that "if every possible effort were made so that the material received by the graduate appeared to be prepared for him individually," a response rate of at least ninety percent could be obtained.[15]

Snelling tested his hypothesis by highly personalizing the cover letters mailed with the questionnaires. A good quality

Figure 28-4. Quotations

tions more than three lines long are set off, so they don't need quotation marks. Triple space before the quote, and single space the quote itself. All the typed lines of the quote begin at the paragraph tab setting. (If the quotation itself begins a paragraph, indent the first line an additional four spaces.) The quote is also indented from the right margin—at least five spaces. Any words you decide to omit should be replaced by three spaced periods (. . .) Triple space after the quote. The finished product should look like Figure 28-4.

WORD DIVISION

Because your paper will look much better if you don't go over the one-inch right margin, you'll have to divide some words at the ends of lines. But you must divide the words correctly, that is, between syllables. Check a dictionary. There are some exceptions: don't carry over one-letter syllables or endings like "ing," "tion," or "ed," and don't divide proper names. Divide hyphenated words where the hyphen is. If it's impossible to divide a word at the end of the line, or if the right margin is getting cluttered with too many hyphens, it's better to carry the whole word over to the next line than to go too far into the right margin.

PAGE NUMBERS

Number pages consecutively throughout the body of the paper with Arabic numerals (1, 2, 3). The first page of the text isn't numbered, but it counts as page 1, so numbering will start with page 2. Pages with major headings are numbered at the bottom margin in the center. Other pages are numbered in the upper right hand corner of the page, one inch from the right margin, on the fifth single space from the top. No punctuation. Text begins a double space below the page number.

CORRECTIONS

Use the typewriter eraser to remove only minor typographical errors. If the error is too big to be erased (extra or omitted words, for example), you should re-type that page. Never "strike over" a mistake with its correction without erasing the mistake first, and never correct a mistake with a pen or pencil.

If the error is small enough to erase without messing up the page, carefully erase the error, *leaving the paper in the typewriter*. It's hard to realign the type once you've taken the paper out, so proofread each page carefully before you take it out.

Now, armed with bond paper, a guide sheet, eraser, ruler, and instructions, you should be ready to type the first page of your paper. Follow the

directions for typing a major heading to get started on the first page. Refer to
the other directions as you need to. Good luck. (And don't forget to come
back and find out how to type what goes before and after the text.)

TYPING THE FOOTNOTE SECTION, BIBLIOGRAPHY, TITLE PAGE, AND TABLE OF CONTENTS

The preliminary pages—those that come before the body of the paper—are
the title page and the table of contents. Since you can't do the table of
contents until you've done everything else (so you'll know what the page
numbers are), you'll want to type first the parts that come after the text, the
footnote section and the bibliography.

FOOTNOTE SECTION

If you put your footnotes at the bottoms of the pages, you deserve a break.
Go to *Bibliography*.

If you saved them for a separate footnote section, do this:

Start a new major section. Treat the title of the section, "Footnotes," like a
major heading. The footnotes are typed in numerical order the same way for
this section as they would be at the bottom of the page: the first line indented
to the paragraph tab, the footnote number half a space up, single spaced,
double spaced between. Keep numbering the pages consecutively (don't start
all over with "1") and use the same rules for placement. See Figure 28-5.

BIBLIOGRAPHY

The bibliography is another major section, right after the text if you don't
have a footnote section, right after the footnote section if you do. Treat the
heading "Bibliography" as a major heading. As with footnotes, you single
space the entries and double space between them. But the indentation is
different. The first line of each entry starts at the left margin, and all lines
after that are indented four spaces. Set a tab for these "underhung" lines.

Check a reference for bibliography form if you need to. If you have more
than one entry by the same author, don't type her or his name in entries after
the first. Substitute an unbroken line eight spaces long with a period after it.
Page numbering is continuous from the body (or footnote section) through
the bibliography, and the rules for placement are the same. Figure 28-6
shows a typed bibliography.

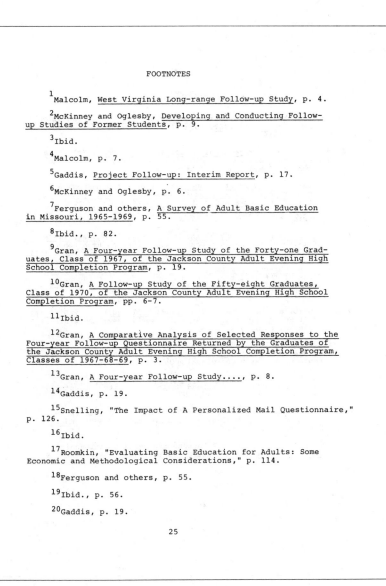

FOOTNOTES

[1] Malcolm, <u>West Virginia Long-range Follow-up Study</u>, p. 4.

[2] McKinney and Oglesby, <u>Developing and Conducting Follow-up Studies of Former Students</u>, p. 9.

[3] Ibid.

[4] Malcolm, p. 7.

[5] Gaddis, <u>Project Follow-up: Interim Report</u>, p. 17.

[6] McKinney and Oglesby, p. 6.

[7] Ferguson and others, <u>A Survey of Adult Basic Education in Missouri, 1965-1969</u>, p. 55.

[8] Ibid., p. 82.

[9] Gran, <u>A Four-year Follow-up Study of the Forty-one Graduates, Class of 1967, of the Jackson County Adult Evening High School Completion Program</u>, p. 19.

[10] Gran, <u>A Follow-up Study of the Fifty-eight Graduates, Class of 1970, of the Jackson County Adult Evening High School Completion Program</u>, pp. 6-7.

[11] Ibid.

[12] Gran, <u>A Comparative Analysis of Selected Responses to the Four-year Follow-up Questionnaire Returned by the Graduates of the Jackson County Adult Evening High School Completion Program, Classes of 1967-68-69</u>, p. 3.

[13] Gran, <u>A Four-year Follow-up Study....</u>, p. 8.

[14] Gaddis, p. 19.

[15] Snelling, "The Impact of A Personalized Mail Questionnaire," p. 126.

[16] Ibid.

[17] Roomkin, "Evaluating Basic Education for Adults: Some Economic and Methodological Considerations," p. 114.

[18] Ferguson and others, p. 55.

[19] Ibid., p. 56.

[20] Gaddis, p. 19.

25

Figure 28-5.   Separate footnote section

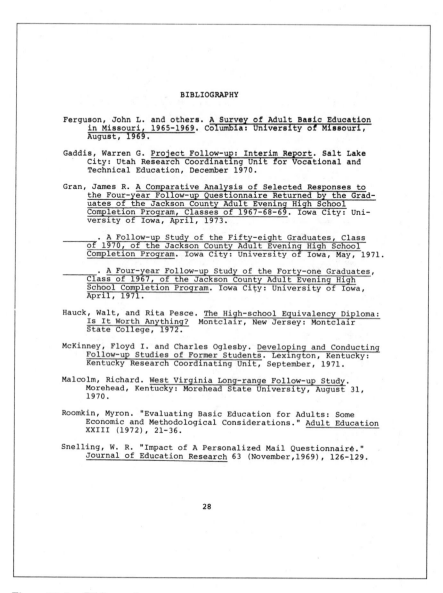

BIBLIOGRAPHY

Ferguson, John L. and others. A Survey of Adult Basic Education in Missouri, 1965-1969. Columbia: University of Missouri, August, 1969.

Gaddis, Warren G. Project Follow-up: Interim Report. Salt Lake City: Utah Research Coordinating Unit for Vocational and Technical Education, December 1970.

Gran, James R. A Comparative Analysis of Selected Responses to the Four-year Follow-up Questionnaire Returned by the Graduates of the Jackson County Adult Evening High School Completion Program, Classes of 1967-68-69. Iowa City: University of Iowa, April, 1973.

_____. A Follow-up Study of the Fifty-eight Graduates, Class of 1970, of the Jackson County Adult Evening High School Completion Program. Iowa City: University of Iowa, May, 1971.

_____. A Four-year Follow-up Study of the Forty-one Graduates, Class of 1967, of the Jackson County Adult Evening High School Completion Program. Iowa City: University of Iowa, April, 1971.

Hauck, Walt, and Rita Pesce. The High-school Equivalency Diploma: Is It Worth Anything? Montclair, New Jersey: Montclair State College, 1972.

McKinney, Floyd I. and Charles Oglesby. Developing and Conducting Follow-up Studies of Former Students. Lexington, Kentucky: Kentucky Research Coordinating Unit, September, 1971.

Malcolm, Richard. West Virginia Long-range Follow-up Study. Morehead, Kentucky: Morehead State University, August 31, 1970.

Roomkin, Myron. "Evaluating Basic Education for Adults: Some Economic and Methodological Considerations." Adult Education XXIII (1972), 21-36.

Snelling, W. R. "Impact of A Personalized Mail Questionnaire." Journal of Education Research 63 (November,1969), 126-129.

28

Figure 28-6.   Bibliography page

TITLE PAGE

The title page has five parts: (1) the title of your paper; (2) the person or group to whom you are presenting the paper; (3) the requirements the paper fulfills; (4) your name; and (5) the date.

Center the title of your paper in all caps, beginning about ten single spaces from the top of the page. Single space and divide the title in inverted pyramid style if it looks too long for one line. ·

Each line on the title page is centered. The vertical spacing arrangement will depend on how much information you have to give in each of those five parts. The finished title page should look balanced, like Figure 28-7.

TABLE OF CONTENTS

Preparing the table of contents involves two major steps: (1) roughing it, gathering the correct information from the typed paper, and (2) typing it in correct form.

ROUGHING IT   First, make a rough draft of the table of contents by listing the following items in the order in which they appear in the typed paper. After the item, put the page number on which it appears:

1. introduction
2. major headings
3. side headings
4. paragraph headings
5. footnote section (if you have one)
6. bibliography

Wording, capitalization, and order should be exactly as they are in the body of the paper. Indent the side headings under each major heading to show their subordination, and indent the paragraph headings under each side heading. If you numbered the chapters or major sections of the paper, those numbers should be given before the chapter title in the table of contents.

TYPING IT   Clear all the old tabs. Set a tab one letter space to the left of the right margin for the column of page numbers. "Table of Contents" is a major heading.

Starting at the left margin, type the first item of the table of contents (usually "Introduction"). Type the page number on which the item appears at the tab you set on the right. For numbers that have more than one digit, you'll have to backspace so the numbers will be even on the right.

Now go back to the end of the typed first entry in the table of contents.

```
                    HOW TO TYPE YOUR PAPER




                    An Instructions Manual
                       Submitted to
                  Dr. Donald H. Cunningham
                Associate Professor of English
                  Morehead State University




                    In Partial Fulfillment
          of the Requirements for Technical Writing 591



                            by
                        Anne Shelby
                      January, 1974
```

Figure 28-7.   Title page

TABLE OF CONTENTS

ii

Figure 28-8.   Table of contents

Starting at the first even number on the typewriter scale below the carriage, type a period on every even-numbered space, stopping at an even-numbered space two or three spaces to the left of the number you typed on the right. (Stop at the same place for each entry.)

Type all the entries this way. Double space between them, and single space any title that is too long for one line. Type major headings against the left margin, side headings two spaces in, and paragraph headings four spaces in. Set tabs for those. A table of contents for this paper would look like Figure 28-8.

When you get the table of contents typed, number the preliminary pages in small Romans, i, ii, iii, iv. The title page is "i" but it doesn't have a number on it, so the first page of the table of contents will be numbered "ii" in the center at the bottom margin.

Now you're finished. But wait. Go away for awhile and let the paper, the typewriter, and your head rest before you all get together for the final round. When you feel up to it, proofread *very carefully* and correct your mistakes *very carefully*. And turn it in.

Suggestions for Further Reading

TITLES AND ABSTRACTS

Aziz, Abdul. "Article Titles as Tools of Communication." *Journal of Technical Writing and Communication* IV (Winter 1974), 19–21.

    Although this article deals specifically with the titles of articles, it contains important information about forming titles for reports, too, since report titles should also contain maximum information in a minimum of words. Titles are good or bad according to how well they describe the content of the piece of writing. Titles should also reflect the audience addressed in the piece of writing.

Cortelyou, Ethaline. "The Abstract of the Technical Report." *Journal of Chemical Education* 32 (October 1955), 532–533.

    There are two types of abstracts: the descriptive (which is much like the table of contents in sentence form) and the informative (which states the problem, method of conducting the investigation, conclusions reached, and recommendations made). Beginning writers might learn to write abstracts by first lifting important sentences from the body of the report and then condensing the passage until it is only 1 to 5 percent the length of the report.

    Abstracts perform three functions: (1) they inform management of the results and recommendations of reports; (2) they allow busy readers to decide whether they should read the entire report; and (3) they allow librarians and information specialists to catalogue material better.

    After writing the abstract, the writer should check the report to make sure that it covers all items mentioned in the abstract, that the abstract and report agree on the relative importance of items of information, and that the abstract is as brief as possible.

Kennedy, Robert A. "Writing Informative Titles for Technical Literature—An Aid to Efficient Informational Retrieval." *IEEE Transactions on Engineering Writing and Speech* EWS-7, No. 1 (March 1964), 4–5.

    A carefully written and accurate title will help readers retrieve and use a report. The title can be considered as a one-sentence abstract that reflects the subject of the report. Among the suggestions for choosing titles, two help writers to avoid common pitfalls: (1) the writer should avoid generally unhelpful phrases such as *A Report on* . . . , *Some Problems Associated with* . . . , and *A Study of* . . . ; (2) the writer should avoid making titles too short or too long. If a title exceeds fourteen or fifteen words, it should be shortened if it can be without severe information loss.

The suggestions for choosing titles can also be used by the writer when choosing headings and subheadings.

INTRODUCTIONS

Hand, Harry E. "Introductions to Formal Reports: A Good Start or a Bad Ending?" *STWP Review* (October 1966), 16–17.

The quality of the introduction often determines how far a person will read in a piece of writing. The first few pages of a formal report are not awkward preliminaries preceding the main event. To judge the adequacy of the information in introductions to thirty master's theses accepted by the graduate school of the Air Force Institute of Technology, the following criteria were used:

1. average length of introduction
2. favorable first impression
3. statement of subject
4. statement of purpose of the report
5. scope
6. plan of development
7. statement of possible importance of study
8. summary of results and conclusions

Although the average length of introductions was 3½ pages, the average length of introductions of the theses rated excellent was significantly higher, whereas the length of those rated poor was significantly lower. If a correlation exists between the length of an introduction and its quality, it may be because an introduction that provides all the necessary items of an introduction will naturally be longer.

Mills, Gordon H. and John A. Walter. "Introductions." *Technical Writing*, 3d ed. Holt, Rinehart and Winston, New York, 1970.

An introduction has several functions: to state the subject, the purpose, the scope, and the plan of development of the report. Sometimes the introduction also explains the value of the subject and summarizes principal findings or conclusions.

INDEX